# ENERGY, ENVIRONMENT, and the ECONOMY

# ENERGY, ENVIRONMENT, and the ECONOMY

**EDITED BY**
SHYAMAL K. MAJUMDAR, Ph.D.
Associate Professor of Biology
Lafayette College
Easton, Pennsylvania 18042

Founded on April 18, 1924

**A Publication Of**
**The Pennsylvania Academy of Science**

Library of Congress Catalog Card No.: 81-82465

ISBN 0-9606670-0-8



For information write to The Pennsylvania Academy of Science, Attention: Dr. S.K. Majumdar, Editor, Department of Biology, Lafayette College, Easton, Pennsylvania 18042.

Printed in the United States of America by

Typehouse
Phillipsburg, New Jersey 08865

# CONTENTS

# FOREWORD

One would be hard put to conjure up a more timely and vitally important subject than "Energy, Environment, and the Economy." The concept of energy is the essence of the universe and our existence in it. The three E's *must* be understood and accommodated if life is to continue on this planet—or at least life as we understand the term. Much has been said about this subject, most of which is from one or another particular viewpoint, something like an "us against them" approach. Recognizing the importance of a rational inquiry into all views, the Pennsylvania Academy of Science has provided a rather unique symposium, and its product can be understood by all.

Let us then read, contemplate and remember the words of General J.F.C. Fuller:

> "Adherence to dogma has destroyed more armies and lost more battles and lives than any other cause in war. No man of fixed opinions can make a good general."

So it is here. We must be willing to keep our minds "open" and search out the best means of solving the most serious problem of our time. I am confident we will.

Mr. Justice John P. Flaherty
*Supreme Court of Pennsylvania*
*and Chairman, Honorary Executive Board,*
*Pennsylvania Academy of Science*

# PREFACE

# Energy, Environment, and the Economy

The book was conceived as a collection of energy and environment related papers presented at the two seminars sponsored by The Pennsylvania Academy of Science. The first seminar bearing the title *Energy, Economy, and the Environment* was held at Duquesne University, Pittsburgh, in November, 1980 and the second one titled *Ecological Implications of Energy Production* was organized during the 57th Annual Meeting of the Academy at Host Corral, Lancaster, in April, 1981. In these two meetings, executives, scientists and officials from industry, university and government examined and discussed various aspects of energy related issues and the U.S. energy future. In addition, experts from various research centers were invited to offer their views on the U.S. energy sources, and to assess the pros and cons of the energy production and its effects on the economy, environment, and the society. Several chapters cover the many aspects of the science of coal. The future of nuclear, solar and energy from non-conventional sources is discussed in other chapters. The concerns resulted from the Three Mile Island nuclear accident are summarized in three papers. Several contributors address the U.S. energy future and energy alternatives.

The editor is grateful to Dr. Bruce Martin, President, PAS; Dr. George C. Shoffstall, Past PAS President, President, NAAS; Sister M. Gabrielle, Past President, PAS; Mr. John P. Flaherty, Justice, Supreme Court of Pennsylvania; Dr. Stephen R. Williams, Seminar Chairman, and Ms. LeEarl Bryant, Seminar Organization Committee Member and Honorary Executive Board, PAS, for helping him to complete the book. Thanks are due to Dr. Robert S. Chase, Head, Department of Biology, Lafayette College for providing the facilities and for his cooperation and understanding. I also extend my sincerest thanks to the contributors for their time and efforts to compile up-to-date information for the book. The competent secretarial assistance of Ilene S. Rabin and Daphna Kilion is acknowledged.

The publication is timely, and the Academy hopes that the book will provide a useful source of information to all concerned citizens including those who actively work in the fields of energy and environment as well as to administrators and regulators.

Shyamal K. Majumdar
*Editor*
September, 1981

# ACKNOWLEDGMENTS

Publication of this book was aided by contributions from:

Commonwealth of Pennsylvania, and

Fisher Charitable Trust, Pittsburgh, Pennsylvania
Gulf Oil Foundation, Pittsburgh, Pennsylvania
Mobay Chemical Corporation, Pittsburgh, Pennsylvania
PPG Industries — Chemicals, Pittsburgh, Pennsylvania
Pennsylvania Power and Light Company, Allentown, Pennsylvania
Rockwell International Corporation, Pittsburgh, Pennsylvania
Westinghouse Electric Corporation, Pittsburgh, Pennsylvania
United States Steel Corporation, Pittsburgh, Pennsylvania

# OFFICERS OF THE PENNSYLVANIA ACADEMY OF SCIENCE

**BRUCE D. MARTIN,** President; Acting Vice President, School of Pharmacy, Duquesne University, Pittsburgh, Pennsylvania 15219

**STANLEY J. ZAGORSKI,** President-Elect; Department of Biology, Gannon University, Perry Square, Erie, Pennsylvania 16541

**SISTER M. GABRIELLE MAZE,** Immediate Past President; Grove and McRobert Road, Pittsburgh, Pennsylvania 15234

**PAUL B. GRIESACKER,** Treasurer; Department of Physics, Gannon University, Perry Square, Erie, Pennsylvania 16541

**SHYAMAL K. MAJUMDAR,** Editor of the Proceedings; Department of Biology, Lafayette College, Easton, Pennsylvania 18042

**DAVID S. OSTROVSKY,** Secretary; Department of Biology, Millersville State College, Millersville, Pennsylvania 17551

**MICHAEL L. BUCHOLTZ,** Assistant Treasurer; Department of Chemistry, Gannon University, Perry Square, Erie, Pennsylvania 16541

**GEORGE C. SHOFFSTALL,** Executive Secretary, Governor's Liaison, President, National Association of Academies of Science, Director of Education and Organizational Development, The Western Pennsylvania Hospital, Pittsburgh, Pennsylvania 15224

**KURT C. SCHREIBER,** Editor of the Newsletter and Director, Science Talent Search

**THOMAS H. KNEPP,** Historian

**CLARENCE MYERS,** Director of Junior Academy

**JUSTICE JOHN P. FLAHERTY,** Chairman, Honorary Executive Committee, PAS; Justice, Supreme Court of Pennsylvania

**MALCOLM KORACH,** Chairman, Advisory Council, PAS; PPG Industries

**GEORGE C. BOONE,** Membership Chairman

**SHERMAN S. HENDRIX,** Research Grants

**DONALD ZAPPA,** Industrial Relations

# CONTRIBUTORS

**F.J. Brenner,** *Chapter 9.* Professor of Biology, Biology Department, Grove City College, Grove City, Pennsylvania 16127

**H.M. Carlson,** *Chapter 18.* Professor Emeritus of Mechanical Engineering, Lafayette College, Easton, Pennsylvania 18042

**R.B. Domermuth,** *Chapter 8.* Environmental Scientist, Pennsylvania Power & Light Co., 2 North 9th Street, Allentown, Pennsylvania 18101

**H.K. Elo,** *Chapter 17.* Lehigh Forming Co., Inc., P.O. Box 799, Easton, Pennsylvania 18042

**M.K. Goldhaber,** *Chapter 14.* Pennsylvania Department of Health

**A.N. Goldhaber,** *Chapter 6.* Administrative Assistant, Office of the Lieutenant Governor, Commonwealth of Pennsylvania, 200 Main Capitol Building, Harrisburg, Pennsylvania 17120

**P. Hollenbeck,** *Chapter 19.* General Public Utilities, Radiological Technical Support Group, Three Mile Island, Middletown, Pennsylvania 17057

**P.S. Houts,** *Chapter 14.* Associate Professor, The Milton S. Hershey Medical Center, Pennsylvania State University, College of Medicine, Hershey, Pennsylvania 17033

**G.W. Leney,** *Chapter 5.* Regional Geologist, Grand Junction Office, U.S. Department of Energy, P.O. Box 2567, Grand Junction, Colorado 81501

**J. McAfee,** *Chapter 1.* Chairman of the Board & Chief Executive Officer, Gulf Oil Corp., Pittsburgh, Pennsylvania 15219

**D.L. Miller,** *Chapter 16.* Chairman of the American Civilization Program, Department of History, Lafayette College, Easton, Pennsylvania 18042

**J.P. Miller,** *Chapter 10.* Associate Professor of Civil Engineering, University of Pittsburgh, Pittsburgh, Pennsylvania 15261

**E.W. Miller,** *Chapter 12.* Professor of Geography, Department of Geography, 302 Walker Building, Pennsylvania State University, University Park, Pennsylvania 16802

**J.M. Nicholson,** *Chapter 3.* Special Assistant to the Administrator, United States Environmental Protection Agency, Washington, D.C. 20460

**C. Roman,** *Chapter 11.* Roman Resources & Development Corp., 2206 Chew Street, Allentown, Pennsylvania 18104

**R.E. Sharpless,** *Chapter 16.* Associate Professor, Department of History, Lafayette College, Easton, Pennsylvania 18042

**W.F. Skinner,** *Chapter 8.* Project Scientist, Pennsylvania Power & Light Company, 2 North Ninth Street, Allentown, Pennsylvania 18101

**J.W. Smith,** *Chapter 4.* Assistant to the President, United Steelworkers of America, 5 Gateway Center, Pittsburgh, Pennsylvania 15222

**R. Leo Smith,** *Chapter 7.* Professor, Division of Forestry, West Virginia University, Morgantown, West Virginia 26506

**J.E. Tarpinian,** *Chapter 19.* Bechtel Northern Corp., P.O. Box 72, Middletown, Pennsylvania 17057

**J.J. Taylor,** *Chapter 2.* Vice President and General Manager, Water Reactor Division, Westinghouse Electric Corp., Pittsburgh, Pennsylvania 15222

**G.K. Tokuhata,** *Chapter 13.* Director, Division of Epidemiological Research, Bureau of Epidemiology and Disease Prevention, Pennsylvania Department of Health, Box 90, Room 1013, Harrisburg, Pennsylvania 17108

**D.E. Zappa,** *Chapter 15.* President, Vector Corporation, 3700 Butler Street, Pittsburgh, Pennsylvania 15201

# Energy, Environment, and the Economy

Seminar I: **Energy, Economy, and the Environment**

**Organization Committee:**

DR. STEPHEN R. WILLIAMS, Chairperson
The Pennsylvania State University
At New Kensington

MS. LeEARL BRYANT
Honorary Executive Board, PAS
Rockwell International

JUSTICE JOHN P. FLAHERTY
Chairman, Honorary Executive Board, PAS
Justice, Supreme Court of Pennsylvania

SISTER. M. GABRIELLE MAZE
Past President, PAS

DR. BRUCE D. MARTIN
President, PAS
Duquesne University

DR. GEORGE C. SHOFFSTALL
Past President, PAS
President-Elect, National Association of
Academies of Science
The Pennsylvania State University

Seminar II: **Ecological Implications of Energy Production**

DR. FRED. J. BRENNER, CHAIRMAN
Grove City College

# Pennsylvania Academy of Science President's Message

After many months of preparation by the membership of the Academy, the symposium on Energy, Economy, and the Environment was presented to the public on Saturday, November 8, 1980 at Duquesne University in Pittsburgh. Those who attended the presentations in person or even watched the reporting of the day's activities on statewide television news, sensed the excitement of what they were hearing and seeing.

From the Academy's viewpoint, this was the major public event of the year, prepared as the culmination of activities starting over two years ago. The Academy owes a great debt of gratitude to Justice John P. Flaherty of the Supreme Court of Pennsylvania, who chairs the Honorary Executive Board of the Academy. This Board is composed of executives from major industries who have pledged support to the Academy in promoting the use of the scientific method to solve problems of the Commonwealth of Pennsylvania. Membership in the Honorary Executive Board includes: Ms. LeEarl Bryant, Rockwell International; The Honorable Richard S. Caliguiri, Mayor of Pittsburgh; Dr. J.G. Cremonese, Fisher Scientific Corp.; Sr. M. Gabrielle, Past President PAS; Dr. Calhoun L.H. Howard, Mobay Chemical Corp.; Dr. Malcolm Korach, PPG Industries, Inc.; Dr. U. Merten, Gulf Research and Development; Dr. Harold Paxton, U.S. Steel Corp.; Dr. George C. Shoffstall, Past-President PAS; Ms. Cecile Springer, Westinghouse; Dr. John A. Tanner, Pullman Swindell.

As early as 1978, the Board began to consider ways in which the Academy could better serve the needs of Pennsylvania. While the Academy membership is diverse in its scientific disciplines, the Academy as a group had never made its unusual gathering of talent noticeable to the citizens of Pennsylvania nor even the government. It was the planning of the Honorary Executive Board which gave rise to a major symposium on problems of our area. The Board refined its thoughts during 1979 and early 1980 to give us the title of the symposium which helps to bring to light the goals of the Academy.

Following the direction of the Honorary Board, a committee of members was appointed to carry through with the procedural aspects of the event. Justice Flaherty continued with his assistance and was the major factor in the Academy receiving the help of the distinguished panel of speakers. The Academy extends its gratitude to him for the enthusiastic support. Chairing the organization committee was Dr. Stephen R. Williams, Pennsylvania State University at New Kensington, who conducted much of the detail of preparation for the arrival of our speakers and the publication of their talks. Ms. LeEarl Bryant of Rockwell International organized the printing and distribution of the brochures, and Dr. George C. Shoffstall of Western Pennsylvania Hospital, Past President of PAS and President of NAAS assisted your President with on site details. The Editor of the Proceedings, Dr. Shyamal K. Majumdar, Lafayette College, has been responsible for the preparation of all the manuscripts delivered at the symposium for publication. The media relations divisions of Duquesne University and of Gulf Oil are thanked for their assistance with publicity.

Therefore, the production of a symposium is no small exercise. Many, many of our members have been involved. A number of them have not been mentioned here since they are too numerous to list but nevertheless, they have played a supporting role in these efforts. I personally thank them all for their help in spreading the word, welcoming our visitors and generally assisting in the promotion of the goals of the Academy.

The speeches of the day, several of which represent policy statements of the sponsoring organizations, are to be published. The contents of the talks are most important to the future development and use of energy within the context of our local as well as national economy and the maintenance of ecological standards. These ideas for the future and the problems facing our industry were presented to the public in an impartial atmosphere of scientific inquiry. In turn, the Academy pledges its support to help solve these problems and to assist with development of energy sources in the years ahead. The published proceedings represent invaluable documents and a resource for Pennsylvania as we proceed to reach the solutions desired. The nation also will see the efforts made in Pennsylvania and will recognize this state for the leader that it is in developing our resources for the welfare of the public.

As President of the Academy, I extend my personal thanks to all the participants and organizers of the symposium and recommend to all the reading of the special edition of the Proceedings which contains the manuscripts of the activities of the day. This becomes a companion volume to accompany the April, 1980 symposium, the Aftermath of TMI, and both become materials for use in future academy ventures.

Dr. Bruce D. Martin
*President*

# Introductory Remarks

## Introduction to the Seminar:
## Energy, Environment, and the Economy
## Duquesne University
## Pittsburgh, Pennsylvania
## November 8, 1980

Permit me to extend to this distinguished gathering of warm *Caed Mile Filte,* which in the Irish language is "One Hundred Thousand Welcomes." This symposium is both timely and unique, bringing together informed and prominent leaders of industry, government and academia to express viewpoints from their respective involvement in this most important subject. We are honored to have all of you with us today.

The Pennsylvania Academy of Science is composed of many disciplines and is not structured to represent any particular viewpoint, but rather to serve as a vehicle for the advancement of science and the community. We hope that by providing this symposium we will have furthered this objective.

The Mayor of our city, Richard Caliguiri, serves with me on the Honorary Executive Board of the Academy, and has been involved in the Academy since our days at Taylor-Alderdice High School during the 1940's. This seminar was, in part, motivated by him and the subject was, to a great part, his idea. I now introduce to you Board Member of the Academy and Mayor of the City of Pittsburgh, Richard Caliguiri.

Mr. Justice John P. Flaherty
*Supreme Court of Pennsylvania*
*Chairman, Honorary Executive Board,*
*Pennsylvania Academy of Science*

# INTRODUCTION TO THE SEMINAR ENERGY, ECONOMY, AND THE ENVIRONMENT

It is a tribute to me to be able to introduce this seminar on behalf of the City of Pittsburgh. I think to have such a seminar in our city—Energy, Economy, and the Environment—you have the three "E's" here, and I would like to add one more and that is Enlightenment. What we need when we are discussing these three subject matters is enlightenment. We have some fine speakers at this seminar who will hopefully enlighten us as to what we have to do on these subjects.

I have just recently returned from Yugoslavia and had the opportunity to see them erecting a nuclear power plant which they anticipate will provide a city with at least 20% of its energy. One of the things they discussed with me and others was the public opinion following the Three Mile Island nuclear accident. This is Yugoslavia's first nuclear power plant, and they are planning nine more nuclear reactors throughout their country. They are having no problems with switching to nuclear energy, but obviously they want to make sure that the plants will be safe and that their energy needs will, in fact, be supplied by the plants.

This type of symposium, with the exchange of ideas and views, can better enlighten us to all the activities connected with energy, our environment, and the one subject that is always so complicated to me, the economy. I would again like to state that it is a tribute to the City of Pittsburgh to have such a seminar, and I would like to express my appreciation and to give my thanks to the Pennsylvania Academy of Science, particularly Sister Gabrielle Maze (past-President of the PAS) for her devotion to the Academy.

The City of Pittsburgh has been very active in moving from basic steel, which is still an important part of our life, to leading in the movement toward research and development across a broad spectrum of fields. Pittsburgh can be a capital of this country in research and development, and seminars like this keep us moving in that direction. It has been my pleasure to introduce the seminar and welcome the participants to the City of Pittsburgh. Thank you so very much.

Richard Caliguiri
*Mayor of the City of Pittsburgh and Honorary Executive Board of the Pennsylvania Academy of Science*

*Chapter One*

# The U.S. Energy Future: Realizing Our Potential

**Jerry McAfee**
Chairman of the Board
&
Chief Executive Officer
GULF OIL CORPORATION
Gulf Building
439 Seventh Avenue
Pittsburgh, Pa. 15219

Jerry McAfee was elected Chairman, Chief Executive Officer, and a Director of Gulf Oil Corporation in January 1976. He had been President and Chief Executive Officer of Gulf Oil Canada Limited since September 1969.

Dr. McAfee served for three years before coming to Canada as a Senior Vice President of Gulf Oil Corporation and of Gulf Eastern Company, responsible for coordinating Gulf's activities in Europe, Africa and the Middle East.

Dr. McAfee earned his doctorate in Chemical Engineering from the Massachusetts Institute of Technology. Jerry McAfee has held numerous prominent positions: he was National President of the American Institute of Chemical Engineers in 1960, and he was one of three U.S. representatives to the Permanent Council of the World Petroleum Congress from 1955-1964.

Mayor Caliguiri, Justice Flaherty, Dr. Martin, members and guests of the Pennsylvania Academy of Science.

I'm delighted that the Academy selected energy as today's topic. It's people like you — leaders in science, business, and government — who will help shape our energy future.

It's also quite fitting that an energy dialogue take place in the Commonwealth of Pennsylvania. Important chapters of the American energy story — were written, are being written, and will be written — in Pennsylvania. The first U.S. oil well was drilled here. And new technologies may make our state once again a significant oil-and-gas-producer. In fact, the past three years have seen over 5,000 wells drilled here.

Pennsylvania has been a path-finder in other forms of energy also. The first nuclear power plant was built at Shippingsport. And the name Pennsylvania is synonymous with coal. All over the state, private industry and academic institutions have placed a priority on basic and applied research on energy.

Pennsylvania is a microcosm of our nation. The U.S. — and Pennsylvania — have vast energy resources. Our country — and our state — have outstanding researchers and technicians. Our country — and our state — have concerned government officials. Our country — and our state — have the potential to solve our energy problems, to make our energy future a bright one.

In energy matters, America does not have to be held hostage — permanently — by foreign oil. Instead, we can choose what kind of energy future we want. In the 1980's we can choose to cut oil imports in half. We can choose to change our economy from one that depends primarily on oil and gas to one that is based on many energy sources. We can choose to make maximum use of the domestic fuels we have in good supply — coal, uranium, solar, geothermal, oil shale, tar sands, fusion, and biomass. We can choose to put aside energy dependence . . . and to achieve energy security.

The future is born in the present. That's why I'm optimistic. I'm encouraged by current trends in energy development. Very encouraged! In spite of the excesses of campaign rhetoric and in some instances in spite of the government — instead of with its help — we have much to be encouraged about. Let's cite a few examples of justifications for this encouragement.

Record amounts of money — public and private — are flowing into energy research and development. The petroleum industry's R & D expenditures are up by one-third since 1979. The Department of Energy's 1981 R & D budget exceeds 5 billion dollars.

Past research efforts are already bearing fruit. For instance, enhanced oil recovery techniques designed to squeeze more oil out of older fields have added about 385,000 barrels per day to U.S. oil production. Also, we have made a real beginning in the development of so-called synthetic fuels — that is, substitutes for petroleum.

In the effort to extract oil from shale, we've recently seen a tremendous stimulation of activity. In fact, some experts see synthetics from oil shale as even now being cost-competitive with foreign oil. That feature puts oil shale in the forefront as a substitute for conventional petroleum.

Our company is a 50-percent partner in an oil shale project in northwestern Colorado. We and our partner estimate that we've got five billion barrels of recoverable oil in place in our "C-a Tract." We're currently speeding up the testing of both underground and aboveground recovery processes. Besides Gulf, about 30 other companies or organizations are experimenting with a variety of technologies to extract oil from shale.

In spite of the current Canadian governmental impediments, we continue to be optimistic about eventually increasing our production of synthetic crude oil from tar sands. Gulf's affiliate in Canada is a participant in the 100,000 barrel-per-day Syncrude project in Alberta. The technology developed in Canada may be useful in developing the smaller but significant tar sands deposits of the U.S.

Another positive sign is the upsurge in conservation and more efficient use of energy. So far this year, gasoline consumption is down 7 percent from a year ago, and our total petroleum demand has declined 9 percent since 1979. This drop in consumption is due — partly — to the growing realization that we really do have an energy problem and people across the country are willing and able to make certain adjustments in their life-styles to cope with it.

But we must not overlook the fact that this drop in demand is also a response — a very natural and health response — to the higher fuel prices which now prevail.

Of course, no one relishes the notion of paying more for any commodity — including fuel. The price of fuel is probably the most emotional and the most misunderstood aspect of the whole energy issue. But it's an economic fact of life that higher fuel prices do have positive effects. They increase supply. They reduce demand. They provide the needed impetus for developing and using petroleum substitutes.

Another result of higher prices is the growth in capital available for aggressive efforts in the exploration and production of domestic oil and gas. Unfortunately, the so-called "*windfall profits*" tax, in conjunction with other federal and state taxes and royalties, will take about 85 cents from each additional dollar we receive from the phased decontrol of domestic crude. On the positive side, the industry still gets an additional 15 cents — 15 cents we didn't have before — for energy development.

What's the result of phased decontrol? A surge in exploration and production activities is the most important result. The 1980 average for drilling rigs in operation was almost three times the number operating before the 1973 embargo. The number of crews working in seismic exploration is almost twice the number working in 1973. As a result of the drilling push in recent years, the new oil reserves added last year were larger than they have been in any of the past eight years.

Incidentally, during the next five years we at Gulf plan to invest some 16 billion dollars in exploration and production activities, out of our total budget of some 18 billion dollars . . . all of it energy-related. And about three-quarters of this budget will be spent in North America.

Prospects are also looking up for the direct use of coal. New technologies, such as stack gas scrubbers and fluidized bed combustion, are making coal use more acceptable environmentally. As you doubtlessly know, the U.S. has more than half of the free world's coal resources. A recent survey found that the public most frequently mentions *coal* as an alternative to oil and natural gas. This energy source can play a major role in reducing our dependence on foreign oil. If public support permits, the U.S. can double its coal use by 1990.

In the area of nuclear power, I think Americans have begun to recover from the trauma of Three Mile Island. A recent Harris poll found that more than half the people favored new construction of nuclear plants and more than two-thirds favored the continuation of power plants now in operation. A short while ago, the Nuclear Regulatory Commission granted a full-power operating license for a new atomic plant — the first such permit to be issued in nearly two years. If that forward momentum intensifies, we can triple the contribution of nuclear power by 1990.

Another heartening development is that there seems to be forward movement toward achieving a balance between energy goals and environmental goals.

For instance, the Environmental Protection Agency has recently approved the use of the "bubble concept" for controlling pollution in some industries and utilities. This concept, rather than establishing a rigid formula for handling pollution, allows a company some flexibility in determining how it can best meet pollution standards.

All these trends show that the country has the potential for energy security. However, it won't be easy. To fully realize our energy potential, we've got to work together — as a nation — to overcome several critical obstacles.

*Obstacle No. 1* is the limitation on capital formation.

The nation will have to commit enormous amounts of capital to energy development. It is clear that if the country is to hold imports to their 1978 level of 8.5 million barrels per day or less, we have to hold our domestic oil and gas production at virtually today's levels — or higher levels. To do this, capital expenditures for conventional oil and natural gas exploration and production will have to double to some 40 billion dollars per year by 1990. The easy oil has already been found, so the industry now must drill deeper and in more hostile frontier areas for those more elusive barrels. This costs money — lots of it! In addition, the country will have to make a cumulative investment of 60 to 70 billion additional dollars to have synthetic fuels plants capable of producing one million barrels per day by 1990.

This need for capital is intensified by inflation. The prices of our major purchases — drilling rigs, offshore leases, and construction and technology for synthetic fuels plants and refineries — have gone up much more rapidly than the Consumer Price Index. The long lead time involved with projects like synthetic fuels makes these undertakings even more vulnerable to inflation.

Yet, as we know, the states as well as the federal government are undergoing budgetary pressures and increased constituent demand for services. And the oil companies are a convenient target for increased taxes.

But, we need to remember one thing: It takes energy to create and maintain jobs, to fuel the technology which can minimize pollution, and to sustain the American dream of the Good Life.

The bottom line is that unduly increased energy taxes can lead to reduced domestic energy supplies.

*Obstacle No. 2* is the lack of access to federally-controlled land and off-shore areas.

During the past 15 years, energy development has been gravely hurt by federal policies which either restrict — or downright forbid — access to potentially energy-rich lands and offshore areas. Unfortunately, the trend is toward withdrawing even more of these prospective areas from exploration. And, if I'm reading the political winds correctly — this dangerous trend will continue — unless we are able to convince both the people and the politicians of the nation that it should be reversed in the national interest.

Right now about one-third of all U.S. land is in federal hands. Yet only one-third of this one-third is available for energy development. The situation is especially bad in Alaska, the Western States, and the Outer Continental Shelf — areas which experts believe may hold abundant oil and gas resources. But, by and large, we haven't been allowed even to explore in these areas to determine the real energy potential. As a result, a disproportionate amount of exploratory drilling has been confined to heavily developed and lower potential areas onshore.

Offshore, the access situation is much worse. Less than 4 percent of federally-controlled offshore areas have been opened to energy development.

This absurd situation exists despite the fact that there is every reason to believe that offshore waters contain vast amounts of undiscovered recoverable oil and natural gas.

How can we produce domestic energy if we can't get at it?

*Obstacle No. 3* is the regulatory quagmire.

Like other businesses, the energy industry is being strangled with red tape, restrictions, and delays. For instance, an oil shale developer is now required to comply with 170 regulatory requirements to get a project off the ground. Some of those regulations involve permits which are sequential. In other words, if we don't get permit "C" in the required time, we could lose permits "A" and "B". Talk about Catch-22!

*Obstacle No. 4* is the historic adversary relationship between government and industry.

We need full, open cooperation between government and industry. Anything less can seriously undermine energy development. The nation no longer can afford the luxury of indulging in constant squabbles between the public and private sectors.

*Obstacle No. 5* is the tension between the oil-producing and the oil-consuming nations.

Energy is an international problem — not one restricted to the U.S. or any other country or group of countries. Petroleum is a finite resource. Sooner or later, it will run out. Members of OPEC publicly admit that they, like the U.S., will have to develop petroleum substitutes.

Our world is becoming more and more interdependent. Both the oil producers and the oil consumers must work to bridge the gap between a system based on petroleum and one based on a variety of energy sources.

During this difficult transition period, the oil producers and the oil consumers must learn to communicate better with one another. We must learn to transcend short-sighted self-interest.

The obstacles we face are formidable, but I'm confident we can overcome them. With sound technology, sensible energy policies, wise leadership, and trust, we *can* achieve energy security by 1990.

*Chapter Two*

# Views from Industry on Energy Alternatives

**John J. Taylor**
Vice President and
General Manager
Water Reactor Division
WESTINGHOUSE ELECTRIC CORP.
Pittsburgh, Pa. 15222

John Taylor is Vice President and General Manager of the Water Reactor Division of Westinghouse Electric Corporation. Mr. Taylor has held this post since 1976.

Mr. Taylor has been with Westinghouse since 1950 and served as Senior Scientist and Manager in the Bettis Atomic Laboratory. This Westinghouse Laboratory designed the first nuclear propulsion system for naval submarines and ships, and also designed the first nuclear power for generation of electricity in Shippingport, Pennsylvania.

Mr. Taylor holds a BA Degree in mathematics from St. John's University in New York and an MS Degree in mathematics from Notre Dame University. In 1974 Mr. Taylor was awarded an honorary doctorate from St. John's.

Mr. Taylor serves on the American Industrial Forum's Committee entitled "Three Mile Island Nuclear Plant to Recovery," and recently he was asked to serve on the National Resource Council for the Chamber of Commerce of the United States.

The prospectus which was issued for this seminar contained a line which fascinates me.

It states that success in solving our energy demands will depend on the utilization of the strength of industry, government, and academia—under the watchful eye of environmentalists.

Suddenly, environmentalism took a new dimension for me. I thought of a symbol that appears in F. Scott Fitzgerald's book, *The Great Gatsby.*

In *The Great Gatsby,* you may recall, there is a billboard which features a huge eye. And as you move through the book, the eye seems always to be there . . . watching you. For Fitzgerald, this may have been the eye of God.

I wondered if it should—for me—symbolize the all-seeing environmentalist . . . watching me as I worked my energy equations.

I contest this suggestion. I think it fails to express either the essence, or the universality, of environmentalism.

There is, in fact, a degree of natural environmentalism in everyone. Environmentalism is not a separate, scolding category of human activity.

We must each accept responsibility in this matter. No other formula for environmentalism can assure its success. We must always apply conscience about the environment at the beginning of any energy-related activity . . . throughout that activity . . . and then in retrospect.

That is not to say that environmentalism should not have its dedicated spokespeople. That's an important role. But I suggest strongly that the most effective environmentalists are those who understand—and even participate in—the total process of solving our energy-economy equations. I include myself in that description.

Let me turn, with all that in mind, to the murky waters of energy, economics, and the always inherent question of the environment.

Energy and economics are inseparable concepts. The wind that filled the sails of Columbus was intended to fill the shelves of the West with spices.

Because the availability of energy made some people, and some nations, more wealthy than others, it has also been the root source of both culture and conflict. We have gained the leisure to support poets, and to enjoy their poetry. But we have also gained the wherewithal to create armies and the agonies they bring.

Energy and economics are, in short, a complex combination—a duality no less creative and troublesome than the split within every human personality between good and evil.

I emphasize this basic point because it is not uncommon today to read or hear of energy which is taken to be a total, unquestionable good . . . or as a corrupting force for mankind. Some would have us bring in an era of automation which would make man a mere caretaker of his productive machinery . . . while others would like to return to primitive simplicity.

Neither course makes sense. The measuring cup of energy is the human soul. We yearn for the freedom of body and spirit which the use of external energy brings, while rightly fearing the Frankenstein potential of energy applied merely because we have the technical capability to apply it.

The heart of the energy-economy question is, in short, a matter of balance . . . and if I may be forgiven a pun—balance on the human scale. We will always have some degree of both evil and good intermixed in the formula. We will always have to make choices.

There are nations emerging in the world today, for example, which will inevitably choose to pollute their air and water as they crawl toward the 20th century. We did it becaue we were fleeing Dickensian poverty. They will do it for the same reason.

Environmental conscience is usually thought of as a luxury in the early stages of industrial development. In the decades ahead, we can expect to see a growing conflict among nations over this point . . . with wealthier nations admonishing emerging nations about their moral obligations to mankind. The battles which have been fought here over the environment have been squabbles between rich children. And since only a small portion of the Earth's surface contains industrially developed nations who have come to conscience, we can expect the entire question of environmentalism to take on massive global importance within the lifetimes of our young people. It is not unlikely that wars will be fought someday between contiguous nations with differing environmental standards. Wars have often been fought over access to water; they can as easily be fought over its quality. The good-and-evil duality inherent in every energy-economy formula make this result inevitable. Prometheus was right.

I do not sketch this larger picture merely to tour the horizon. I raise it because it strikes very close to home. Within our very prosperous nation there is also an emerging nation. There always will be. It is the nature of the American experiment. America is an unfinished symphony. It is a process—a method by which have-nots can hope to become haves . . . and the have-nots we will always have with us.

When we discuss America's choices on energy, economics, and environmentalism, we have to face that fact. It is idle to suggest that America need grow no more, or that the pie is big enough to slice right now.

America must grow economically because there are people pushing to climb the stairs. This fact can be expressed in mathematical terms—in ergs and barrels, but its essence is the restless human condition—against which neither arguments nor armies can prevail.

People *will* demand better lives. They *will* insist on opportunity for their children. They will, as they have in the past, raise revolutions to get these things if they must.

A growing America is therefore a politically stable America. Zero growth or anything approaching it is an invitation to economic chaos . . . and after chaos, either anarchy or the iron hand.

To grow economically, we must have more energy. No forecast which inspires any real confidence suggests that we can conserve enough energy to meet the social aspirations of the American people without a steadily growing energy supply.

Once that premise has been accepted—and it is plain that I consider it a necessary premise—we can get down to the specifics of the matter.

The first of those specifics is an umbrella question that is most often answered . . . even though no one seems to be asking it: How much energy will America need for the balance of the century?

I don't know. Neither does anyone else—despite the fact that predictions are blowing around like November leaves.

This is an answer in search of a question because of the global political and economic volatility that raised it in the first place. Who can rationally ponder two decades into the future when the single decade we have just completed has turned the world on its head?

How many ergs can dance on the head of a pin?

The question of how much energy America will need is one that misses the point. A more workable question is: What direction should we be going with energy?

We can't deal with the total needs of the year, 2000. But we can take immediate steps in the right direction here in 1980.

Among those steps in one which has so far been almost totally neglected—and I would like to raise it now for discussion in this Symposium. It is the question of energy flexibility.

We have tied a thousand Lilliputian knots in the past decade—trying to control the energy giant. We have so thoroughly complicated the energy question that most learned discussions on the subject now deal with the tying or untying of single knots . . . while the giant lies inert, unable to serve us. One major example will, I think, make the point . . .

Today, it takes anywhere from 10 to 12 years to build an electric power plant. That's costly, and chancy. It is the main reason why electric utility executives walk around these days with somber expressions. What a job they have—to look ten years ahead when tomorrow another Iraq may move against another Iran. Today, they guess on the high side when they make such estimates, because the cost of guessing on the low side is too great.

A far more rational specific approach would be to reduce the time it takes to build such plants. Five years is plenty for a coal-fired electric power plant. Six years is plenty for a nuclear plant. By shortening these lead times, we can more accurately align the nation's electric power needs with its investment for supply . . . and in the process cut the cost of such plants in half.

It is—I'm sure you realize—the people who pay for power plants. Today's payment to the electric utility buys tomorrow's electric supply. Why we persist in making the supply more costly and imprecise by extending the process of building such plants is hard to explain.

There is no direct relationship between the time it takes today to build a power plant and the safety or environmental acceptability of that plant. Most delays are based on either antagonism or overlapping regulation. We can no longer afford the antagonism, and we ought to rid ourselves of the costly inefficiencies of poorly structured regulation.

We do not need to know how much power will be required in the year, 2000. What we need is the flexibility to move faster and more economically to meet the much more definable needs of the next five or six years. That is one example of an energy knot which needs untying. There are many more. Flexibility in energy planning is a virtue we have too long ignored—and even sinned against.

Now let me discuss a question which is in search of an answer . . . and I will offer a very clearcut answer.

The question concerns direction: What should America's energy mix be? How much of our energy should come from sources such as oil, gas, coal, nuclear, solar, and wind and conservation?

I can answer the latter part of the question easier. We must have as much energy from renewable sources and from conservation as possible.

Renewable energy is not an arguable concept. It makes sense. But the question of when . . . and of economics . . . has not been fully answered. In my company, we have a special gruop working on solar cells. They are not trying to figure out how to design a good photovoltaic cell. That technology is already here—100%. But they *are* trying to work out an automated production line that will make those photovoltaic cells at a price which can compete with conventional central station electric power.

There is still a long way to go. Over the past 5 years the cost of such cells has been reduced from 100 times too costly to only 10 times too costly. That's real progress.

We have groups working on fuel cells, on wind power, and synthetic fuels, and other concepts. All of these people are eager to see their particular technology sweep into popular use. But they all see a long road ahead.

Here, again, there is little point in trying to quantify the precise contribution of such renewable sources during the balance of this century . . . beyond general estimates which can help us structure our thinking.

It does make sense, however, to quantify a goal. As you may have read, the Southern California Edison Company recently announced its intention to achieve as much as one-third of its power generation from renewable sources by the year, 2000 . . . an accomplishment which would make renewable energy almost as important to that company as nuclear energy.

Whatever may be accomplished by such bold initiatives, the major role in the nation's total energy supply must certainly come from more familiar sources. Oil and gas seem certain to supply a steadily smaller percentage of our needs, but will still increase in absolute terms. Coal will do a greater share of the job, as will nuclear energy.

Such general changes in direction are reasonably predictable—even if the exact numbers which accompany them are not.

But I would like to suggest . . . again for the sake of discussion . . . a partisan viewpoint in the matter of direction. I make to you a single, simple proposition . . .

To the degree that an increasing share of our energy is put to work in the form of electricity, we will be able to solve more of the economic and environmental problems asociated with energy.

Electricity has been part of our lives for so long that we tend to think of it as we do air and water. We flick the switch, the lights light, the motor hums, the oven heats. The supply is so reliable that on those rare occasions when the power fails, people tend to have parties. Or looting. And sometimes babies.

The trouble with the familiarity of electricity is that it breeds a lack of understanding of the extraordinary nature of this form of energy.

It's amazing stuff. It will serve at milliwatt levels to operate an electronic computer. Or it can be applied at megawatt levels to run a 10,000 horsepower steel mill motor.

It creates light, heat, torque, and communications. But it's all the same thing —just flowing electrons.

Now consider the fact that this extremely flexible form of energy is created for the most part from primary energy sources which are totally independent of our friends in OPEC. Coal is all American, and so is much of uranium. The uranium we must buy elsewhere comes almost entirely from friendly nations such as Canada.

The moral of this display of partisanship is straightforward: To the degree that we can shift our economy *away* from the use of imported oil . . . and *toward* domestically produced electricity, we can solve some very pressing problems . . .

We can reduce the inflationary economic effects of constantly escalating OPEC prices and the creation of petrodollars.

We can reduce the strategic importance of Mid-East oil, with the ominous potential there for military conflict.

We can restore American pride. We can put money in American pockets.

All of this is not quite as jingoistic as it may sound to some. In fact, a reduction of American dependence on foreign oil will do much to relieve the pressure on the bank accounts of emerging nations, who must now compete in the OPEC marketplace with us for their oil. Our demand drives up their prices. They in turn borrow repeatedly to buy the oil. Someday, that house of debt must tumble.

This Nation, in effect, owes itself and the world a moral obligation to disengage as much as possible from the OPEC entanglement.

A swing to electricity can do much of that task. To quantify the potential would be to guess . . . but let me whet your theoretical appetites a little . . .

Nearly a third of all uses for oil involve the creation of heat. There is no heating job—in the home, the store, or the factory—which cannot be performed as well by electricity.

In an ever widening portion of the Nation, the electric heat pump is becoming economically competitive with oil and gas for heating.

Equipment similar to heat pumps is already available which allows a factory to recapture heat from activities such as welding and use it to heat water or space . . . thereby reducing the dependence of that factory on oil and gas.

Electric industrial furnaces can do the same job that gas-fired furnaces do.

The largest single use of oil is for transportation. That great American love affair with the automobile spurred us to build a Nation which is woven together with roads and superhighways. It will not be easy to extricate ourselves from that web.

But much can be done. Electric-powered streetcars, buses, and trains work just as well as diesel equipment.

We once had electrified interurban systems such as those which ran between Gary, Indiana and Chicago. We should look at such ideas again—even if the tracks run parallel to the interstate, and even if government subsidy would be needed. Such subsidies are no more than a method of rerouting the taxpayer's dollar from a system that will take part of it to the Mid-East into a system that will keep all of it in the domestic economy.

The small electric car can play a similar role for those with limited, specific driving needs. People who drive every day to punchclock jobs can sensibly rely on cars that recharge their batteries on lower cost overnight electricity.

Today, about 8% of the oil used in the United States . . . some 14% of the total that is imported . . . is used directly in electric power plants to create electricity. Electric utilities are working to reduce this use of oil. They should succeed most rapidly in places like New England as planned nuclear plants come on line.

If we had a national policy to do so . . . and if we had a crisis attitude to support that policy, it would be possible to replace as much as half of all oil uses with equally good or better electrical systems. Electricity can—in theory—totally replace the need for the importation of *any* foreign oil.

Why do we not have such a national policy, and such an attitude?

The main obstacle is economics. The bitter bottom line is that it is less costly for individual Americans to use imported oil at the moment than to make such a massive shift to electricity.

Just as we are a National of highways, we are a Nation of oil furnaces—both in the home and in industry. To change those furnaces . . . and to change the distribution systems which feed them . . . would be expensive. And so we are being tempted along the path of potential disaster by the lower short-term cost of imported oil.

We don't need to subsidize electrical equiment to change that situation. We need only accelerate the process of making imported oil more expensive to use. A person, or a company, buy a furnace today . . . knowing that oil will soon be too costly to use . . . will choose an electric furnace. He will, in effect, burn easily available domestic fuels.

There are also environmental considerations to be dealt with in a shift toward greater use of electricity. As always, choices must be made.

The unavoidable fact is that neither electricity nor any other form of energy will serve us without a price. Inevitably, we are driven into a corner which we can escape only by answering the question posed by conscience.

What environmental price will we pay . . . can we pay . . . must we pay for the energy we want—whatever form of energy it may be?

For hydro, we must sacrifice valleys and the animal life and architecture within them.

For coal, we accept sulphur in the air.

For gasoline, we must live with nitrogen oxides. Los Angeles suffered its worst smog in ten years late in September.

For nuclear energy, we must accept slight amounts of radiation release . . . and we must learn to deal with the fear of a phenomenon which . . . however well known and acceptable it may be to its engineers . . . is mysterious to the great masses of people.

And according to last Monday's *Post-Gazette,* we must even face the good-and-evil duality of the wood-buring stove. The paper reported that the Insurance Federation of Pennsylvania has found that carless installation and maintenance of wood burners has resulted in more than $2.5 million in fire losses last year in the State.

How much will be paid for the adequate and secure supplies of energy which we need to meet the nation's growing needs? I can answer such questions for myself. But I cannot answer them for you.

I do propose to you, however, the following simple formula—with the hope that we will enjoy a lively discussion of it this afternoon.

I offer you the premise that the nation must grow economically if it is to fulfill its unending promise to all men. That promise is to give everyone a decent chance . . . . and to stand as an example to the world that there is a way for all men to live together politically in a system that can make such a promise and pay off on it. I will not accept the idea that such a notion is corny. It is not worn out. It is not in error. It is right, and it is as permanent as the heart of man.

I propose in addition that the essence of America's answer to its energy problems should be a massive shift toward the use of the most thoroughly domestic of all energy forms . . . electricity.

I acknowledge that this answer is not perfect. A perfect conscience in such matters leads only to absolute immobility. We must protect our environment to the greatest practical degree—knowing that total protection is not within our reach.

I suggest to you as a vital corollary that a full reassessment of nuclear energy is now overdue. We have enormous benefits . . . economic, social, and environmental . . . to gain from it. Emotional, know-not, think-not reactions to the word "nuclear" are a luxury we can ill afford.

Our greatest risk—whatever direction we may choose to take—is in doing nothing.

We are drifting now. We are taking the course that is easiest for the day immediately ahead. We are sustaining with our shrinking dollars the infrastructure which threatens our standing among nations, our internal order, and our economic and social future.

We much decide which direction to take. I have proposed such a direction for you.

We must decide when to act. And I say to you that we are long overdue for action.

*Chapter Three*

# The Role of the U.S. Environmental Protection Agency in Our Energy Future

**Joan Martin Nicholson**
Special Assistant to
the Administrator
UNITED STATES ENVIRONMENTAL
PROTECTION AGENCY
Washington, D.C.

Joan Martin Nicholson was Assistant to the Administrator for Consumer Affairs for the United States Environmental Protection Agency and responsible for agency-wide consumer programs and policies, as well as all public information, citizen participation and press.

Ms. Nicholson has a B.A. in Political Science and International Studies from Allegheny College. She has designed and directed energy education and conservation programs for the Federal Administration in 13 states and served on President Carter's Energy Task Force. In April 1981, Ms. Nicholson was appointed the United Nations Environment Programmer's representative, Washington, D.C.

I think it would be helpful to us all this morning, Mr. Mayor, Mr. Justice and the Academy of Science to have a chance to explore together how we got where we are. In the most negative sense of the word, how did we as a nation get where we are in terms of energy problems, economic problems, environmental problems? What can we learn from looking at these three topics in a unified way which could be constructive in policy making and in dealing with the relationships between Government, the public, and the corporate world. There is much to be learned reminiscent of the three people on an island. They are told that a huge tidal wave is going to come in the next 24 hours and inundate the island and that they are all doomed. To cheer them a little, they can each have one wish granted. The first woman said she would like to go off and see all the historic monuments in the world; the pyramids, the great wall of China; the fates agree and off she goes to be back in 24 hours. The second person said that he would like to see all the high spots and live it up for 24 hours. After some thought the third person said, "I would like to have the fates bring me all the books that have ever been written on how to live under the sea." That story perhaps reflects the real relationships between energy, environment and economics, and perhaps implies an agenda for ourselves as a nation.

One of the things we do not look at when we begin to talk about energy policy or an economic policy is the fact that land, air and water are the basic life support systems. Whether black or white, male or female, in the North or the South, or in part of the new economic or the old economic order, one breathes a minimum of 12,000 quarts of air every 24 hours; the human body is 60% water, and the saline content of one's blood is the same as that of the oceans. We have evolved as a species, as a part of the natural environment around us; and over millions of years, our system and the food chain that supports us has evolved in harmony so that our species can be sustained. The land, the capacity of air sheds to absorb pollution, and the water sheds to sustain life, are as finite as the capacity of the earth to generate coal, gas and oil. We do not always focus on these resources as finite. The land, air and water are not only finite resources, but they are the fundamental underpinnings to economic health, to community stability and to the capacity of humans to be active in economic production. You do not see many industries or communities in the middle of the Sahara Desert. From this perspective, it becomes apparent that if we begin to evolve policies which relate economic, energy and the environment as a whole, we may have something concrete with which to begin.

The historic capacity of water sheds and air sheds and even land to absorb pollution has worked pretty well in the past. In fact, nature's skill at converting pollutants in essence was a free commodity. When people and their activities over generations of times produced pollution, the carrying capacity of land, air and water actually seemed to have an allowance or capacity to absorb and convert pollutants in a way that fed back into biological systems, and the contamination factor was pretty well balanced. With the tremendous expansion of human knowledge, technology, economic activity, and the demand for consumer goods, we have come to a very peculiar time in the history of our species. We have overproduced and overconsumed beyond the capacity of land, air and water systems to absorb pollution, and it is coming back to haunt us.

Since World War II, we have introduced over 30,000 chemicals from our technological knowledge which are synthetic, which means these chemicals did not evolve naturally from the biological systems. The land, air and water systems simply cannot take them in and convert them to a benign state. It is important conceptually to look at this because, as Dr. McAfee said, we have had a history during the past 20 years of increasingly strident antagonism between environmentalists, public interest groups and corporations. This antagonism may have stemmed from not articulating the problems accurately and may have cost us opportunities to integrate ideas from various interest groups advocating economic, energy or environmental perspective.

Let us look at a very fundamental case, the case of a lake that draws people in the summer for recreation. The value of the land around that lake goes up; people build houses, fish, and swim; the "Seven-Eleven" comes in because they now have a market. The county gets new tax funds because the community is expanding. It is a growing, economic situation. Upwind or upriver somewhere a new activity begins. Maybe it is a government military installation - the corporations do not do all the pollution in the world - or maybe it is a corporation or a municipal sewage facility. Ultimately, pollution ends up in the lake. First, the property values begin to drop because the lake is less desirable and people are not coming there in growing numbers. The drop-off in population in the summer causes the "Seven-Eleven" to fold and eventually people become politically active. They go to the town council asking a bond be passed to clean up the lake and stop pollution. Invariably, they confront a jurisdictional problem as the county line does not encompass the facility doing the pollution.

If it is in the affected jurisdiction, then the county can float a bond. You are now taking money out of the public troth, which comes from the consumer/taxpayer. In essence, to build the "widget" on the lake to clean it up requires energy and public dollars. If that corporation or municipal facility upstream had been able to avoid pollution, or if the bond had gone initially to help them avoid the pollution, the lake may have never been polluted. The destabilizing effect to the lake's economy and the pollution liabilities that may be irreversible, such as kepone in the James River which wiped out a whole fishing industry, would simply not have happened. Our society should look for technology and economic formulas that prevent pollutants from being released into the land, air and water in the first place.

Interestingly, we often use economic formulas to assess the mess. There is this thing called the "G.N.P.," the gross national product. I have wonderful discussions with my friends trained as economists as I am not one. I often have the feeling I have been to the parade and the "economist King" is stark naked. Let us look at G.N.P.; the measurement of goods and services in our society. If a lake does not get polluted because of some regulation, there is no contribution that is visible or measurable to the G.N.P. Let us say the lake did get polluted; we have to float a bond; and we have to hire a contractor. The contractor builds something, and jobs have been created. We now hire a truck to take the sludge out of the lake, (if that's the problem) and haul it away to some land-fill site which may, in fact, contaminate the aquifer it is being put on. If there is an epidemic that breaks out, people go to the doctors. New jobs and doctor bills are considered as positive economic indicators to the G.N.P. I have often said that the Environmental Protection Agency should get its budget determined by the number of lakes or streams that do not get polluted that year in term of the basic economic value of an unpolluted area. Somehow our measuring tools that determine money values are not working.

It shouldn't be a benefit to the gross national product when public debts or public funds are having to be used to clean up problems that should not exist in the first place and where public health may well be jeopardized. These are some of the problems, as you begin to look at the interrelationships between energy, economics and the environment.

If we keep looking at the problems the same old way and using the same old measuring sticks to see how we are fairing as we look to the '80's, our sense of options are locked in. We have to begin to look at these issues differently, and I think we will begin to feel more optimistic about their solutions. President Carter talked about the national malaise; I do not think there is a national malaise. I think there is a national frustration with defining these same old problems the same way and losing our sense of options for the future.

One thing of great concern to me is how we manage knowledge and promote education in the Western world, the academic traditions. We are seeing the glaring effect of that in the EPA as a federal agency. The EPA engineer uses a completelly different vocabulary than employees in research and development. There are also soft and hard scientists, and we have the economists. They all ignore and sabotage each other in the EPA. When they do try to communicate, they do not understand each other. The energy engineer looking at coal mining effects will talk about overburden or nonpoint pollution. The scientist will talk about things unpronounceable. It took me three years to say "epidemiological retrospective longitudinal surveys." Government has locked itself into vocabularies just like doctors, engineers and teachers. We are not communicating in Washington with each other. The main reason is a traditional compartmentalization of knowledge.

As knowledge grows and we become more and more intense in our effort to master knowledge, we compartmentalize it in what I call the vertical plunge, more and more expertise, more and more specialization. There are very few people coming out of academic institutions who do not reflect academic turf problems, who are integrators, who understand relationships such as those between biology and engineering. There may be profound patterns in nature that can lend themselves to either be replicated technologically in design or that we can plug into with as fundamental a profession as architecture. We groan at the energy bills paid to cool our buildings. Yet, we cannot open one window. How many of us look at the positioning of buildings to take advantage of the progression of the sun across the sky during the day, or the cycle of deciduous trees to augment how and what amount of energy we have to use? There must be opportunities for academic institutions and for young people to challenge how we compartmentalize knowledge leading to myoptic fixes.

Why do you have all these regulations Dr. McAfee referred to? The reason regulations are introduced is because private sector mechanisms fail to exist or function or the local mechanisms of government failed to work, failed to exist or were not applied. When you get kepone dumped into the James River and the James crosses state boundaries and you are dealing with aquifer systems, a whole integrated extensive network of water systems becomes evident. You have to begin to find mechanisms that respond and one is a regulatory fix. I have said to my corporate friends, if you do not like the federal fix (and you should not), get off the federal fix and fix it yourselves. Their response is that they cannot, they have the Sherman Antitrust Act, or they do not have jurisdictions, or will be at a competitive disadvantage. If a coal plant spends money for a scrubber and the competitor does not, the competitor has an economic advantage. Why are not some of the trade associations given mandates to look at how they could become the think-tanks for their industries? In addition to the lobbying role that they play, they could look for mechanisms to voluntarily develope standards. The oil industry, in spite of the Sherman Antitrust Act, has been able to standardize pipe fittings that work the world over. They have been able to join funds to transport fuels around the world and lease from each other in transport cargo. If they have been that innovative in terms of the area of their expertise, surely there could be some kind of corporate think-tank developed that looks at these problems anew and tries to solve them before the federal agencies act.

What do you get when the federal agencies introduce a regulation? When the law is being developed in Congress, public interest groups, environmentalists, and corporations all try to be active in the development of that legislation. I will be a little facetious; at 9:00 A.M. the public interest gruop comes in and says this subsection of the law should be strengthened. "We are really going to hold these guys." At 10:00 A.M. the corporate lobbyist comes in and says to add a comma here and a parenthesis there. By the time the regulation comes out of Congress and comes over to the EPA, it is already in a language that nobody can decipher. Many times there are deadlines with those laws, so the agency throws a group of well-meaning people together, and they start to grind it out to beat the deadline. Everybody is automatically an expert, but are they? I see people working on these regulations who have doctorate degrees, but few, if any, have been on a production line. None have been underground looking at long wall operations in coal mines; none of them have been on site in offshore drilling. These are very complex technological systems. Is a technological prescription always the best? Sometimes it is, sometimes it is not. The corporate world, however, has not been trying to work with public interest groups to seek innovative economic incentives or other innovations that could be developed by the private sector to respond unilaterally to solving environmental problems.

What I suggest (and it is not always supported by EPA) colleagues, is that one year before any public servant makes a regulation (sometimes they are not public servants, sometimes they are bureaucrats, and there is a big difference) the EPA puts out a document called declaration of intent to regulate. The document would state the environmental problem and the adverse effects it is having. Furthermore, it would state the goal so that public safety is protected or the environmental life support systems are protected and then request proposal for solution. For one year, the EPA would do nothing. It might be the best money the public ever spent to have the federal agency do nothing for a whole year. If public interest leaders, corporate leaders, other federal agencies or any other interested party developed and prepared concept papers or position papers during that year, we would have had some kind of a public dialogue rather than a 30-day announcement. At the minimum, the person writing the regulation would get an education and perhaps the regulation might be recognized as not the best approach. Maybe a regulation was not necessary because a corporate entity through its trade association had decided to sign a contract to achieve the EPA goal by a set date and have it binding before the courts.

What I am concerned about and one reason I was very pleased to be invited here today, is that rather than people skulking about on a win-lose agenda, which we as a society cannot afford any more, we stop looking for win-lose agendas. There were days I closed my eyes in the administrator's conference room and in corporate board rooms and listened. You could be in a locker room before the game of the Pittsburgh Steelers, "we are going to get those guys; we are going to nail them to the wall." If the corporations "win," we are all going to lose. If some environmentalists win, we all lose. We must begin to look for win-win agendas. The reason the corporations are in such trouble environmentally is they have a legacy which they did to themselves; such as Three Mile Island, Love Canal and the vast soil erosion due to agrobusiness and mining practices. We have lost more topsoil since 1978 than we did during the entire dust bowl, and yet food is one of our most critical exports in the world balance of payments. Look at what can only be called the rape of the land before the big lumber companies got organized as to how to stabilize soil and timber correctly. Too often corporate behavior has generated terrible liabilities for the public that only courts, government agencies and public interest groups combined could change. Corporations and government are often organized alike. Too often regulations end up as being bizarre and counterproductive. Most corporate organizations have what they call profit centers, governments have program units. Corporations look at two things; how much is produced in a given amount of time, and how much is marketed in a given amount of time. These figures determine corporate borrowing capacity in the money market and whether the stockholder wants to invest. Government meets Congressional and election schedules on a short-term basis. The results determine how long a Chief Executive Officer (C.E.O.) or Senator holds his job. Do you know that in the U.S. a C.E.O.'s position in the United States is shorter now than a four-year cabinet officer appointment? That is how tremendously competitive the marketplace is and the political arena is similar. Now, when the ability to attract capital or votes is predicated on short term performances, it is very hard for corporate or government employees who say: "You know I think we should defer this product decision for five years. I am amazed that in ten years corporations and government has done as much as they have, on environmental problems because the institutional evolution of corporations and government certainly has not lent itself to behave in a way that would preclude a lot of these pollution/regulatory problems. Yet, almost every respectable corporation (everyone in *Fortune* 500) has senior level environmental officers. Companies like 3M are looking at the interrelationships between economics, energy and environment and the three "E" measurement for every policy they review.

We have got to begin to look at ways for government, public interest groups and corporations to work together. If I had been a C.E.O. in 1970, I would have hired the head of every single environmental group I could get my hands on, and I would have given them all fat contracts and said you are the experts, "help us." "You help develop institutional mechanisms and the political processes to solve these problems so that people keep their jobs, and we stay viable in the world economy." It would have been a lot cheaper than going through the courts with all those environmental leaders. I think that is the kind of approach we are going to have to look at much more closely. We must think in terms of synthesizing knowledge and integrating policies. To conclude, the more problems you fix in Pittsburgh environmentally, the more you can fix in Pennsylvania environmentally, the less you will see of the EPA and the fewer there will have to be of us, and the lower your federal taxes will be, and everyone will be a winner.

*Chapter Four*

# Some Economic Aspects of the Energy Movement from Labor

**James W. Smith**
Assistant to the President
UNITED STEELWORKERS
OF AMERICA
Five Gateway Center
Pittsburgh, Pa. 15222

Mr. Smith has served as Assistant to the President of the United Steelworkers of American since 1971. Mr. Smith was President of the Local Union in Houston from 1948 to 1953. He was Staff Representative for Texas, Oklahoma, and Arkansas from 1953 to 1957 and from 1970 to 1971.

James Smith has a B.A. in Economics from the University of Texas at Austin and an M.A. in Economics from the University of Houston. He is Acting Research Director for United Steelworkers and has organized his own Energy Symposium with a grant from the Department of Energy. From 1975 to 1976 Mr. Smith served as President of the Western Pennsylvania Chapter of Americans for Energy Independence.

Our Union has been concerned with the energy problem for a number of years. I would say that most of us involved in the leadership of the union, like most of the people in the country, seriously became aware of an energy problem following the 1973/1974 embargo by the Arab Nations of oil deliveries to the United States, and the resulting jump in the price of petroleum during 1974, and the economic events that our nation suffered then.

We conceive the business of the labor movement to be to speak to the human needs of our society, the ethical problems, if you will, of our culture from the standpoint of ethics as the definition of values in human life; and we view our role in our society as being one of looking beyond next year's balance sheet, or next quarter's balance sheet, as most business firms must do, of attempting to anticipate the human problems of perhaps not only next year, but the next generation.

If there are significant ways in which our points of view in organized labor, differ from the points of view of those in management who represent the owners of industry, I would say it is in this respect: that we attempt to have, and we should have, a longer time horizon; and secondly, our responsibility is more to human needs than the dollar needs, and that's not to deprecate the responsibility of the earlier speakers or anyone else in management of business firms.

The business firm is the particular institution that our society has deliberately selected through its political processes and economic processes to deliver goods and services that it values. We could have selected other instruments—we didn't . . . historically and presently.

Frequently, I am involved in activities that also involve environmentalists, and I have become more acquainted with some of them in Pittsburgh and their representatives in Washington. I would say that one of our problems in viewing all of these inter-actions between, not energy and the environment, but industrial processes and the environment generally, economic issues and the environment, is the newness of the environmental movement.

The environmental movement today reminds me of the labor movement in the 1940's and the late '30's when all of us were still reacting, I think, out of our anger and our anguish at the great depression, and issues were black and white. There were villains and there were heroes. We weren't just adversaries, we were enemies.

Well, we grew out of that. We learned to accept the fact that there is a partnership between management and labor, between owners and workers. In our society—we cannot divorce ourselves, that we must handle our relationships in such a way that we can meet tomorrow and negotiate and solve problems. We must accept the continuing existence of one another.

I think there are many in the environmental movement who still have ahead of them the acceptance of these fundamental relationships, though intellectually they accept it. Emotionally they are still operating on the win-lose propositions. By the same token, there are many, many people in management who still haven't emotionally accepted the permanence of the environmental movement and who dream of the day that they will win the ultimate political victory that will wash the environmental movement out of their hair, but we'll see. They have come as close to it now as I think they ever will, and we'll see if that happens. I don't think it ever will.

The environmental movement has enough reason for its existence to be a permanent aspect of our lives; I think it will be, and I think it has a contribution to make.

Speaking to our own concerns, working people in an industrial society need energy in order to live in the way that we have become accustomed to live in our culture, to move from one place to another, to cook our food, to heat our homes, to do all the other things we do as consumers of energy. We also need energy to work. Very little work is done by physical effort—it is done by controlling the use of some artificial energy source, if you will, and causing that energy to do work, directing the work done by that energy.

Just as traditionally in the low production societies of the past and of other parts of the world, where few people direct the labor of many, today one worker in the United States or Western Europe or Japan directs the labor of energy systems that produce the goods and a good many of the services that we use. So we have a stake as workers in the energy system of our culture, and we have a stake in that energy being supplied at dependable and stable prices.

When those prices become unstable—when the supply becomes undependable—we pay for it with unemployment as well as inflation. It is precisely that experience that we are having now, and it is caused precisely by energy shortage and the economic policy response of our particular economic and political system to that energy shortage.

I said earlier that I became concerned, our union became concerned, about the energy problems of six years ago. Actually I was told before then, earlier in the 1970's, by scientists that I respected who were in major positions, senior positions in the American Oil industry, something that I never thought of before. I knew, I guess intellectually, that oil was a finite resource, but it never occurred to me that that meant anything to my life or would during my life. But I was told that the United States was running out of petroleum and that the world would start running out of petroleum in the sense of its petroleum production peaking sometime in the 1990's. I began to wonder then how it would happen.

I remember telling one of my sons who had just turned 16 about this information that had been given to me and that I had confidence in the judgment and knowledge of the person who had given it to me. He reflected on that throughout our dinner that evening and then announced that he thought it was pretty rotten—that the grownups had used up the petroleum just about the time he was ready to go get his driver's license. I assured him we hadn't used it all up yet, but I don't think it helped his feelings any, and nothing that has happened since then on the subject has helped his feelings very much either.

He just bought his first car the other day—the first one he bought with his own money I should say—and he's begun to pay his own gasoline bills, you know, and all that. He thinks that we have abused his generation, and perhaps we have. Perhaps we have.

How will it happen? Will we live normal lives essentially uninterrupted or unchanged until one morning we get up and the oil is all gone? — No, we all know that. What will be the nature of the change? Well, it is, at least at this point that a little training in economics tends to help in anticipating what happens.

As a particular raw material becomes scarcer and scarcer, its price rises, and as its price rises, substitute goods to the extent they are available begin to be used in place of that commodity. Users of that commodity begin to substitute other activities for the things they've traditionally done with that commodity. As the price gets up to certain levels, we just stop doing some of the things that we previously did that required the use of that particular commodity.

What does this mean in the case of petroleum? Well, essentially it means that we're going to use more of other fuels. We could, as the previous speaker said do most of the things we now do with automobiles driven by gasoline engines, with vehicles driven by electrical engines and supplied with electricity by plugging them in the garage every evening to a recharging system. We could make fuel to run internal combustion engines out of coal, or out of corn or out of garbage or out of a lot of other things.

Particularly, we could make it out of oil shale mined in certain Western parts of the United States where we have potentially more fuel embedded in that oil shale than the petroleum in Saudi Arabia. We could make it out of tar sands. We could make it out of certain heavy oils that are almost tars that are found in Venezuela and certain other places in the world.

So, as the prices go up we will shift to these alternative fuels, and as prices go up, we'll quit moving around so much. Also, we'll learn to stay in one place. We will quit doing some things that we traditionally did because of the higher cost. We'll go back to public transportation much more than we have traditionally used public transportation, because the higher prices will force us to.

Now if all this change were to happen in a very gradual way—gradually prices rising, gradually shifting to other materials, gradually going to public transportation—the problems generated for labor, the ethical problems, if you will, and the unemployment problems, would not be severe. Workers would adjust as everyone else would adjust.

Unfortunately, that is not the way it's going to be, and if the record tells us anything, the record makes it very clear that the free-market system will not function in that way, with respect to conversion of our lives from one system to another. That is the essential meaning of the present depression.

No one really knows how a free-market system in energy would function because we've never had one. There's no point in kidding ourselves now that the elections are all over and there is no need for rhetoric on these subjects. We can be frank. The American oil industry has certainly never been a competitive free-market system since John D. Rockefeller recognized the social evils of competition in petroleum and created the Standard Oil Company in something like 1901.

There have been other oil companies admitted to the club, since then, but on the condition that they behave themselves in proper and non-competitive ways. I don't say that to belittle them. . . the oil industry or its managers. Pure competition in energy led, even in some aspects, to ridiculous things like 10¢ a barrel oil in my native State of Texas when there was pure competition in drilling and producing oil—just in that aspect, not refining or marketing.

Pure competition led to utter wastefulness of resources. Pure competition led to every single residence down certain streets in certain small towns in East Texas having its own oil well in its own backyard; all sucking oil out of the same pool that maybe one well for each hundred of those could have provided, so let's don't kid ourselves. Regulation has been the way of life for the oil industry.

The State of Texas used to regulate the industry, regulate production of oil in the state in order to see to a reasonable price return for the investment, and that system worked reasonably well. It brought the price of oil up from 10¢ a barrel to $2.50 to $3.00, and it stayed there until the Arabs decided to regulate the oil market instead of letting the Texas State Government regulate the price of oil.

They would have regulated it, and they have—not the Arabs, alone, but OPEC as an institution. They discovered their market power in 1974 and they quadrupled the price of oil from $3.00 a barrel to about $12.00. Following the Iranian revolution they discovered their market power again in 1978 and '79 and they tripled the price of oil from about $13.00 a barrel to between $35.00 and $40.00. After the Saudi Arabian revoltuion they will discover their market power again, and even us middle-aged folks can anticipate $100.00 a barrel oil within our productive lifetimes. These things are coming.

We are depending on the most unstable political system in the world—in the Middle East—to supply us with this necessary ingredient of our culture, that our cultures currently won't function without. Not just our culture, but the Japanese and Northern Europe cultures as well, will continue not to function without petroleum until we make the conversions that I was speaking about earlier, and which earlier speakers have mentioned.

Because of our reliance on essentially free-market systems to make the adjustments in 1975, we had a rate of inflation then that got all the way up to 12%. We had a rate of unemployment then that got all the way up to almost 10% for a few months of measured unemployment, and we in labor believe that the true unemployment is probably 4 to 5 percentage points higher than the measured unemployment.

Therefore, in 1979 we vigorously opposed the legislation in Congress to completely decontrol domestic oil prices, or de-control rapidly. I won't say completely, because they are still not completely decontrolled, but the de-controlled process has been greatly speeded up. Most of our domestic oil is coming out of wells which were drilled 5-10-15 years ago; much of it coming out of pools of oil that were discovered and were profitable at $3.00 a barrel.

We were quite concerned about the impact on the economy if that $3.00 a barrel oil suddenly became not just $12.00 and $13.00 a barrel oil, but $30.00 and $40.00 a barrel oil, because of the world market price dictated by OPEC. We felt there would be economic shock waves sent through the economy that would cause unemployment with massive numbers of people if this was allowed to happen.

In spite of our best efforts, however, the de-control of oil was adopted as a national policy in 1979, primarily because President Carter felt that all of his efforts to get programs in place to control energy had been largely ineffective because of opposition in Congress and from the public, and that the only remaining way to conserve was through high prices.

So, the President and the White House pushed the legislation. We lost, and we are now getting exactly what we expected. Automobile sales of domestic automobiles dropped some 30 and 40% earlier this year, partly because the American public wanted more fuel efficiency at these new prices of gasoline—gasoline prices almost doubled within a 12-month period.

The American automobile factories were not geared up with equipment to produce fuel-efficient automobiles in that volume, and it will take them a few years to get so geared up. Consequently, their sales have plummeted; their orders for steel have plummeted; they have 30% of the American automobile industry employees, and not only those in the automobile factories, but those in all the supplying facilities, laid off; and we have enormous unemployment in the steel industry today—25% at its worst, and it's still around 20%.

These are human beings; these are not just numbers—these are men and women with bills to pay, families to feed, children to raise and educate and so on. Our society is experiencing these shocks, and these shocks, of course, postponed for an even longer period our efforts to eliminate poverty in this country.

Over about a 10-year period, we succeeded in reducing the percentage of people in this country living in the measured poverty economy, from some 21% of the population of the United States, down to about 11%; but since 1975, movement on that war stopped. We haven't lost much ground until this year. I think we have lost ground this year.

We haven't gained any ground in the last five years. We won't, as long as our economy is stopping and going, is up and down, as our major industries are experiencing these kind of shocks that I've just described in autos and steel.

When we lost the fight to Congress over rapid de-control, the administration and Congress did give us a bone, and the bone they gave us was the Windfall Profits Tax. They said, look, we know that we're going to increase the value of that oil in the ground that the oil industry of the United States already owns, by hundreds of billions of dollars, and, therefore, we will enact a tax that will claim some 80 billion of those dollars over a period of 10 years, and we'll spend that money largely on efforts to conserve and to convert to synthetic fuels and to other energy systems.

Well, we were in favor of that in organized labor, if we were going to have to take the beating on de-control, because we think it is important. We don't think the free market is going to give us the amount of oil shale production, the amount of synthetic fuel production from coal and from corn; we don't think it's going to give us the amount of conversion, or solar energy, or nuclear energy that we're going to need. Any sort of program of this nature that would spend some revenue in that way is certainly worthwhile, so we were happy to get that bone, although it was not what we really wanted. What we really wanted was a much more gradual transition on the price break.

Now, the rhetoric of the Reagan campaign has been to repeal the Windfall Profit Tax and to reduce or eliminate public expenditures to assist in the development of the synthetic fuels industry or oil conversion and to turn the whole energy problem back over to the "free market," which we are convinced is not a free market, never has been, and probably shouldn't ever be.

Of all markets, the least free is the electrical energy market, which is a totally regulated public utility type market. We believe that that rhetoric, if carried out, would be a prescription for continued and even greater economic instability in the 1980's and 1990's than we have experienced in '75 and '80. Therefore, we will oppose repeal of the Windfall Profits Tax. We will oppose reduction of expenditures on synthetic fuels and conservation.

In general, I would say we believe that every possible means of saving energy should be encouraged by public policy and private advocacy. Secondly, that every possible means of developing substitute, alternative energy sources should be done and should be done as quickly as they can be done safely.

We're not frightened about nuclear energy, most of us. Some people in labor are. In general, I would say those of us in labor that have studied the problem, including United Steel Workers President McBride, who was a member of the Kemeny Commission which studied the Three Mile Island incident, are convinced that nuclear energy can be produced safely.

We are not necessarily convinced that it is being so produced now. Insofar as there is a problem in producing it safely, one of the major problems, of course, is the same problem that exists in doing anything else safely. Doing things safely in industrial processes does not necessarily mean doing it the cheapest possible way or the most profitable possible way. We would advocate, for your consideration that a possible solution to that problem might be to build needed nuclear power plants at public expenditure and to build them according to every possible, or necessary, or useful safety regulation and system, and that the power from them be sold to private power distribution companies by public agencies comparable to the Tennessee Valley Authority. We don't think power, necessarily, has to be produced for profit; although particular municipalities that prefer profit-making public utilities to publicly-operated utilities for distribution of it, certainly should continue to have that right.

Finally, I would say this: In energy, as in many other things that we do in our culture today, we need more light and less heat. The only way we see to get it is to look at what's being done in Germany and Japan and certain other Northern European countries where there has developed more of a social contract between management and labor and other public interest groups in the society, and through which they have been able to do a much more effective job of policy planning.

I'm not advocating total economic planning. I am an advocate, however, of policy planning, in the sense that government policies need to be coordinated much more than they are today to achieve socially desirable objectives. One of our major problems is that each separate agency Congress has created, or the state has created, or the county has created, operates almost in a vacuum pursuing its particular legislative mandate without regard for the other mandates given to the other agencies of government. So we have situations that I, myself, have experienced repeatedly in the steel industry, and I know they also go on in the electric power industry, the oil industry and others in which one agency is telling an industry or a company to do this and another rules, "you must not do" the same thing.

It is not just the number of regulations, it is the contradictory nature of many of the results of applying some of these regulations, and the totally uncoordinated approach that our culture is currently taking towards these problems. The environmental movement and the EPA, which is the creature of the environmental movement, takes one approach; the Justice Department and the anti-trust laws that it enforces and the believers in that, that are of their constituency, take another approach, our trade policy takes still another approach. Each is made for a particular purpose, and we need to coordinate.

No one is sitting down in the government and saying—how much electrical energy are we going to need 20 years out—30 years out—and what mix of environmental policy, trade policy, anti-trust policy, etc. will best get us there. Nobody is saying that about the steel industry or any other industry, and I think that all of us in our very special interest groups need to look for ways that we can meet together to promote systems of policy planning and compromise, if you will, of our various opinions and interests to achieve that.

*Chapter Five*

# The U.S. Department of Energy and Some Perspectives on Uranium and Coal

**George W. Leney**
Regional Geologist
Grand Junction Office
U.S. DEPARTMENT OF ENERGY
P.O. Box 2567
Grand Junction, CO 81501

Mr. George W. Leney's present position is that of Regional Geologist for the Grand Junction Office of the U.S. Department of Energy, P.O. Box 2567, Grand Junction, CO 81501.

Mr. Leney holds a Master of Science degree in Electrical Engineering and a Master of Arts degree in Geology from the University of Michigan. He was previously employed as a private consultant, and is a former Chief Geologist for H.K. Porter Company, Inc. in Pittsburgh, former Chief Geophysicist for The Hanna Mining Company, Cleveland, Ohio, as well as having worked for major oil companies, and the Geological Survey of Canada. He has a wide variety of experience in exploration and development for energy minerals and fuels, industrial minerals, iron ore, and base and precious metals.

Mr. Leney is a member of the Society of Economic Geologists, the Society of Exploration Geophysicists, the American Institute of Mining Engineers, and the Geological Society of America. He was a recipient of the Robert Peele Memorial Award of AIME, and is the author of a number of scientific and technical papers. He is a registered Professional Geological Engineer in Pennsylvania, and is a biographee in Who's Who in Engineering.

The subject of energy and the impact of new developments on the United States economy and the environment we live in is a matter of serious concern to the nation. In order to understand the role of the U.S. Department of Energy, it is necessary to first outline the view of the future, as expressed in various statements by Department administrators and staff, and then describe how it is organized to obtain these goals. We will then proceed to some specifics on coal and uranium, since these industries are of major concern in Pennsylvania. As the Chairman has indicated, my role is that of Regional Geologist for the Grand Junction, Colorado office, which administers the National Uranium Resource Evaluation Program, and I would like to briefly describe our work as an example of one type of activity of DOE, and how the actions of local and State governments can impact on an energy industry. The Department of Energy embraces a variety of policy making, regulatory, and management functions; however, since my office has a purely technical responsibility, and this is a scientific organization, we will address ourselves to that point of view.

The nation's energy problem did not arise suddenly. It is the result of a dependence on foreigh petroleum that has been growing for several decades. By the same token, a solution to our energy problem will not occur suddenly, or as the result of a single technological breakthrough. DOE's program and organizational structure recognizes this fact. It is designed to advance a multiplicity of technologies over a period extending well into the next century. Projections in the National Energy Plan II, for the year 2000, taking the high price case, foresee domestic production of all forms of energy rising by 70% from 62 to 106 quadrillion BTU equivalents, and imports dropping from 16 to 10-12 quads. Most of the drop in imports would come from oil. These have already declined from a 1977 peak of 8.6 million barrels per day, to 7.9 in 1979, and to 6.8 million by May of 1980, with a goal of reducing them to 4.5 million bpd within the next decade. Coal output would increase by 22 quads, from 17 to 39, more than double the current production, nuclear energy from 3 to 17 quads, and domestic oil from 20 to 22 quads. Natural gas would continue constant at 19, and solar, hydroelectric, geothermal, and other technologies would increase from 3 to 10, or an even more recent projection of 18 quads. These are high case figures. Other projections suggest we could limit our consumption to present levels of about 80 quads, or even decrease it to 60, by conservation alone, while continuing economic growth. Conservation would reduce the high case figures by varying amounts, and permit some flexibility in individual growth rates.

These are long-range goals. In some cases, the technologies, such as nuclear fusion or large scale generation of electricity from solar energy, are not yet in place. They will not make a significant contribution in the near future. For the intermediate term, synthetic fuels, for which the technology is available, but which require the development of a whole new industry, will make a contribution by the end of the decade. For the short term, we must rely on the expanded use of coal, in part by the conversion of existing oil-fired utility generating plants, on stimulating oil and gas production, on expanded use of conventional nuclear technology, and most significantly, on conservation. In the end, energy independence will depend on science and technology providing alternatives to imported oil at a cost we can afford, in an environment we can accept and transmit to our heirs.

The Department of Energy was established on October 1, 1977, and is now just over three years old. Its essential mission is to manage national energy policy in ways that *reduce oil imports.* The DOE organization currently includes six major energy-related program elements. A seventh program office is devoted to special nuclear research in behalf of the Defense Department.

The *Office of Fossil Energy* promotes programs designed to increase domestic production of all fossil fuels, especially to facilitate greater reliance on coal and coal-derived fuels. Research and demonstration projects are underway to improve exploration and recovery technologies for domestic petroleum and natural gas. They include tertiary oil recovery methods and ways to produce gas from unconventional sources, such as Devonian shales and tight sands, and from geopressured aquifers and coal seams.

The *Office of Conservation and Solar Energy* directs programs designed to improve energy efficiency and reduce consumption in buildings, transportation, and in agricultural and process heating, and to apply solar technology in buildings, agriculture, and industry. They are supporting things like development of new types of automotive engines, some with non-petroleum based fuels, electric vehicles, and new home heating-cooling systems.

The *Office of Nuclear Energy* is charged with improving technologies using uranium and other fissionable resources. Policy stresses the key role of the present generation of fission energy systems — the light water reactor operating in a once-through fuel cycle mode.

The *Office of Energy Research* supports programs in the basic sciences, including work in high energy physics, materials science, engineering, biology, mathematics, and earth sciences. Some subjects are special metal alloys for use in combustors and fusion reactors, and new chemical processes for more efficient production of hydrogen from sea water. The Satellite Power System envisions capture of solar energy in space.

The *Office of Resource Applications* is responsible for policies and programs to increase domestic supplies of petroleum and natural gas, coal, and uranium. It seeks to develop alternate supplies from oil shale, municipal wastes, tertiary oil recovery, geothermal energy, and small hydroelectric systems. It sponsors research in electric energy systems and oversees marketing from five Federal Power Administrations. Its work includes the Grand Junction Office and the National Uranium Resource Evaluation Program. Some of these functions will be assimilated by the new Synthetic Fuels Corporation.

The *Office of Environment* is responsible for our agency compliance with the National Environmental Policy Act, and manages an important environmental research and development program. They prepare "Environmental Readiness Documents" to identify hazards that may be associated with new energy technologies and help "build in" compatibility with environmental goals. They are supporting research on the "greenhouse" effect, and are working with EPA and others on such things as "acid" rain.

Three semi-autonomous organizations are also included within the overall DOE management structure. These are the *Energy Information Administration,* which gathers and publishes data on energy reserves, production, consumption, and demand. They analyze the data to understand energy trends and conduct field audits to verify accuracy. The *Federal Energy Regulatory Commission* has responsibility for setting rates and charges for transportation and sale of electricity and natural gas, licensing hydroelectric projects, and establishing charges for pipeline transportation of oil. The *Economic Regulatory Administration* administers programs other than those of FERC. These include oil pricing, allocation, and import programs. They also administer programs that mandate conversion of oil and gas fired utility and industrial facilities to coal.

With that overview of DOE and its long-range goals, we would like to take a brief look at the specific subject of coal, which is of special interest in Pennsylvania. Coal is the one fuel we have plenty of, and the industry's capacity to produce is as much as 20% under utilized. The principal focus of the coal program is to create additional demand. The program consists of five parts: reducing utility consumption of oil by replacing it with coal and other fuels, an acceleration of coal leasing on Federal lands, a program to increase coal exports to 80 million tons per year by the end of the decade, the first phase of a synthetic fuels industry which would produce up to two million barrels of liquid and gaseous fuels per day by 1990, now the responsibility of the Synthetic Fuels Corporation, and investments in coal research and development that will exceed one billion dollars in 1981.

In the field of fuel conversion, DOE has already issued orders for conversions involving 175,000 barrels of oil per day, that could translate into 15 million tons of coal demand in the relatively short term. On leasing, DOE has recommended to the Department of the Interior that sufficient additional acreage be leased to ultimately make available about 8 billion tons of reserves, about 50% more than at present. Many current small leases should be consolidated or exchanged with others to form tracts capable of being worked efficiently. It is hoped to structure leasing in such a way that some synfuel plants can be built near the mining sites. A recently formed National Coal Export Task Force will set targets for increasing steam coal exports. International trade in steam coal is expected to expand rapidly, and U.S. exports should increase from 65 million tons last year to 80-120 million tons by the end of the decade.

The synfuels program, which had been under DOE, is expected to generate demand for 200,000,000 tons of coal by 1990. By July, 1980, DOE had announced grants of $200 million for 110 projects to proceed with feasibility studies or plan production systems. Some 970 business organizatins and State and local governments applied for grants, indicating an unusually high interest in swift synfuels development.

Research and development money has been appropriated for atmospheric fluidized-bed combustion systems capable of burning high sulfur coal, for improving energy conversion efficiency in fuel cells, and magnetohydrodynamic systems, for a coal-oil mixture program, advanced coal conversion and combustion research, direct and indirect coal liquefaction, coal liquids refining, in-situ gasification, and coal preparation. Many of these activities are being carried out in the Pittsburgh Energy Technology Center.

It is expected that all of these coal programs will generate not fewer than 59,000 direct jobs in the coal industry and 295,000 jobs in mining communities, not including jobs from the synfuels program. In all cases, the *Office of Environment* will be studying the short and long range impact to ensure that developments are environmentally sound.

The other specific subject I would like to address, is the role of nuclear energy, which is of interest in western Pennsylvania because of the level of involvement in the reactor business. For the current position of DOE, I would like to refer to a statement by Mr. John C. Sawhill, former Deputy Secretary of Energy. Mr. Sawhill was recently nominated as Chairman of the Synthetic Fuels Corporation. "The responsibility of the Federal Government for the future of nuclear energy is clear. We must insure that nuclear power remains a viable option to those electric utility executives who are weighing decisions about investing in additional electric generating capacity. We are doing that in a number of ways:"

"We are first and foremost, taking steps to assure that nuclear power plants are built and operated safely. Reactor safety is the most critical element of any program which seeks to restore public confidence in nuclear energy. Second, we are assuring that nuclear waste materials, whether produced by the military, or by the private sector, are handled and disposed of safely. Third, we are reducing the risk of nuclear weapons proliferation. Fourth, we are improving the regulatory process and expediting nuclear plant licensing and siting. And, finally, we are carrying out a research and development program that will build a strong technical base for the current generation of light water reactors, and for breeder reactors, should they be needed in the next century."

The paper concludes "Only when *all* energy options have been considered, and the need for, and cost effectiveness of new nuclear capacity has been established, both in the minds of management and consumers, will a decision to "go nuclear" gain the broad base of public support necessary for the program to succeed."

The particular aspect of the nuclear energy program that is the responsibility of the Grand Junction Office, is the assessment of the resources of natural uranium to fuel the current generation of reactors. The National Uranium Resource Evaluation program was started in 1974, when the rising price of uranium, coupled with projections for a large growth in demand, made it appear that U.S. resources might be insufficient. A primary objective of the program was to improve the assessment as an aid in long range planning. The program has been carried out by DOE's Grand Junction Office, its prime contractor, the Bendix Field Engineering Corporation, the U.S. Geological Survey, and numerous state geological surveys, universities, and private firms as sub-contractors. This is an appropriate time to discuss it because our "Assessment Report on Uranium in the United States of America" was just released in Grand Junction, Colorado, on October 22, 1980. The report represents a major milestone, but study of uranium resources and revisions of the assessments will continue.

The NURE program used a three pronged approach to the determination of resources. These involved an airborne radiometric reconnaissance, which covered essentially the entire U.S. along flight lines 3-6 miles apart, recording spectrographic data related to concentrations of uranium, thorium, and potassium. A second phase involved collection of samples of steam sediments, stream waters, and ground water, at a spacing of one sample per five square miles, over about two-thirds of the country. Based on these, there was a geologic follow-up in about 135 geographic quadrangles, each 1° latitude by 2° longitude, in the most promising areas. Quadrangle study involved examination of all known uranium occurrences, studies of the geology, and follow-up of indications from airborne and hydrogeochemical data. Resources were assessed by "geologic analogy" to known mining districts, with estimates of probability ranges for four variables, which results in a distribution of probabilities for various amounts of $U_3O_8$. A total of 646 individual areas were assessed. Resources are classified as either reserves, or probable, possible, or speculative resources, and identified in categories by the forward cost of production per pound of $U_3O_8$. At the most widely used figure of $50 per pound, forward cost, resources in all categories are considered to total about 3.5 million tons of $U_3O_8$, with a 5% chance they are as low as 2.8 million tons, or as high as 4.3 million. In the October, 1980 Uranium Assessment Report, "It is concluded that there are sufficient domestic resources at the $50 forward-cost level to meet fuel requirements for currently projected long term nuclear power demand through 2020. However, this will require economic incentives for sustained exploration efforts, as well as success in discovering and developing the potential uranium resources."

As an aside to the discussion of resources, we would like to point out that, as with many programs, there is a scientific fallout from NURE, that may be of benefit in other areas. We have been breaking new ground in the area of resource assessment, which is a developing science, where techniques have application to other commodities for which long range planning is essential. We have also generated a wealth of data from airborne and hydrogeochemical studies. We are currently experimenting with EPA on use of our radiometrics to establish regional baselines and backgrounds for normal radiation exposure rates. The spectral data can be analyzed for indications of other elements.

We have mentioned the necessity for sustained exploration to locate and develop the projected uranium resources. Within the last year, we have seen a movement to ban uranium exploration and mining in several promising areas in the east, where some of these resources might be found. The Vermont legislature passed amendments to its statutes last year, that would require a specific act of legislation of the General Assembly, showing it was in the public interest, to explore for, or mine, fissionable materials. The ban resulted from publicity about a company investigation of a promising prospect. Two townships in New Jersey have recently passed ordinances banning uranium mining in areas where Exxon, Chevron, and Sohio were reported to be working. There are two bills in the legislature to extend the bans state wide. The popularity of these bans is based on a public perception of uranium mining as environmentally destructive, and dangerous to public health.

Uranium is not a rare element in the earth's crust. It is present in rocks of all types, in amounts of 2-4 parts per million, which makes it more common than things like antimony, mercury, or silver, and about one third as common as lead. In downtown Pittsburgh, we are sitting on about as much natural uranium, to a depth of one mile, as was produced and consumed in all the reactors in the U.S. last year. All of this uranium is decaying normally, and releasing its daughter products to the environment. It is a natural part of the world we live in.

Exploration does not disturb that environment, and any concerns can easily be relieved by requiring cementing of exploration drill holes. Bans on exploration simply foreclose the nuclear option, or insure the same foreigh dependency we are trying to avoid with oil. Mining and milling obviously do disturb the environment, and can create local health hazards, but a generalized ban confiscates the property of the mineral owner. Licensing and the issue of permits are appropriately a compromise, decided case by case, on the basis of individual merit, according to the rights of all parties, the protection of the public, and the highest priority use of the areas involved. Leaving resources in the ground is a luxury of an affluent society, and any discussion which does not consider the potential economic benefits of a mining industry, on the physical health and well being of the community, is incomplete, and this is equally true of any commodity, or type of energy development we might consider. Automatic public acceptance of hypothetical worst cases, posed by critics, is as unjustified as an overly optimistic dismissal of potential problems. I am delighted that the Pennsylvania Academy of Sciences chose this subject, and this particular title for a seminar, because it recognizes the interrelationships, and the responsibility of the scientific community to gather the facts on any energy alternative, to analyze them, and to see they are fairly presented to the public.

We have tried to outline some of the programs and policies of DOE that might be of interest to the membership of the Pennsylvania Academy of Science. DOE is a vast organization with an enormous array of research, demonstration, management, and policy making programs. These programs are designed to advance the single goal of reducing our nation's dependence on imported petroleum, and achieving this in an environmentally acceptable manner for the ultimate improvement of economic conditions and the quality of our lives. It is a necessary goal that the nation can achieve, will achieve, and is already making good progress on, with the determination of everyone, and the support of organizations such as yours.

# *Chapter Six*

# What We Are Doing in Coal


**A. Nathaniel Goldhaber**
Administrative Assistant
Office of the Lieutenant Governor
COMMONWEALTH OF PENNSYLVANIA
200 Main Capitol Bulding
Harrisburg, PA 17120


A. Nathaniel Goldhaber is currently Special Assistant to the Lieutenant Governor of Pennsylvania. He served as interim director of the Governor's Energy Council, assisted in system design for the distribution of federal energy grants, and coordinated civil defense planning during the Three Mile Island crisis. Mr. Goldhaber was a former Executive Vice President of Maharishi International University. Mr. Goldhaber received his B.S. and M.A. from Maharishi International University and did post-graduate work at the University of California, Berkeley. He has written several articles in the field of energy.

I was most interested that our previous speaker, Mr. Smith, chose as a central theme for his presentation the issue of de-control of oil prices. According to Mr. Smith, the congressional decision to de-control oil is a significant factor in, if not the direct cause of, our national, economic instability.

It is vital that government stay out of the market place, particularly the energy market place. Governmental interference in energy pricing, even when motivated by the most beneficient of intentions, spells disaster. I use the same example of Mr. Smith's — the control of oil prices — as a most persuasive argument for federal *non*intervention in energy.

In the early 1970's, when OPEC first started to feel its muscle, Congress began to worry about the impact of increasing oil prices on our nation's poor, our industries, and our traditionally petroleum intensive ways of life. Price controlled, cheap, domestic oil supplies were economically mixed through the "Entitlement Program" with expensive imported oil supplies. The market price for oil to end users remained relatively low, far lower than the international price of oil. Incorrect economic signals were passed on to our automobile industry, our industrial sector, and to home owners. Detroit didn't build more energy efficient cars, electrical utilities and manufacturing industries continued to convert from coal to low-priced oil for their major energy supply, and the people of Pennsylvania built homes with inadequate insulation and oil-fired furnaces. Business leaders and home owners didn't get the right information because oil was artificially cheap in the United States as distinct from our allies in Europe and in Asia.

In addition, this Congressional meddling in the market place of oil created a counterincentive to the expansion of the oil industry in the United States. Oil executives said, "I've got this stuff under the ground, why the heck should I take it out when I'm getting so much less than the world market price." We experienced a domestic decline in the production of oil; and this, coupled with a tremendous expansion in demand, forced the United States to turn to the international market to make up needed supply. As the imported proportion of our national oil mix increased, the price of oil to the consumer continued to go up. Simultaneously, our nation began to be vulnerable to international energy blackmail. Finally, Congress caught on and moved for the gradual deregulation of oil prices. The effect was a jump in oil prices at a time roughly coincident with the revolutionary events in Iran.

The tremendous increase in oil prices caused a rippling effect through the national market place. Auto workers, steel workers, and coal miners all experienced significant lay-offs. The automobiles were too inefficient to compete with European and Japanese imports. The market for American steel in the automobile industry rapidly declined, and the metallurgical coal needed to make that steel became unnecessary. The result—unemployment, economic dislocation, and terrible inflation.

The orientals have a word for what we now call the laws of cause and effect. That word is "*karma,*" and it is said that the ways of *karma* are inscrutable. That means that you never know all the consequences of an action. The market is just like that. If you manipulate the market, there will be unpredicted consequences. No one can know all the implications of manipulation. The market should be left alone, especially by government. What started out as a defense against economic dislocation for our poor, our industrial sector, our economy as a whole, completely backfired.

What then is the appropriate role for government in energy? As you know, Lieutenant Governor Scranton is the chief energy officer of Pennsylvania, the Chairman of the Governor's Energy Council. Recently, in addition to overall energy coordination, we have been asked by the Governor to examine the issues of coal from an energy and economic development standpoint. These issues have been addressed cogently in a staff draft document entitled "Pennsylvania Energy Choices," an energy policy plan for Pennsylvania released last May for public review. Since then, the Governor's Energy Council has received thousands of pages of testimony, which are being compiled, collated, and incorporated into the final draft of Pennsylvania's energy policy to be released early next year.

The three main goals of the energy policy are: (1) the maximum economic efficiency in the use of energy; (2) a reliable energy supply system at the least cost; and (3) increase the use of indigenous Pennsylvania energy resources. Now, what do these goals mean, especially within the context of least government control and greatest market place freedom?

The first goal is economic efficiency in the use of energy. That means, to the extent possible, remove government controls that artificially influence the price of energy in the market place so that the private sector can act accordingly. Where there is waste of expensive energy resources, the economic incentives of conservation will, without manipulation, bring about savings.

A reliable energy supply at the least cost—what does that mean? That means that free market forces—not bureaucrat planners—should design the energy supply systems. The only role for government in this area is to assure a free market through antitrust regulation.

And the third goal is to increase the use of indigenous resources—that means dig more coal. That means that we should eliminate unnecessary and consolidate necessary mining and environmental regulations; fund studies to discover new, efficient ways to reduce coal emissions; and create a useful coordination point for important coal development projects.

Let me give you some specific examples of what we've been doing in coal as an illustration of how government can assist the private sector in energy development without interfering in the free market place. One way to do this is to set a good example. In Pennsylvania we are now retrofitting major state-owned office buildings with efficient coal heating systems. In addition, we are installing two or three innovative coal burners that virtually eliminate sulfur emissions, while simultaneously using a lower-quality, less expenisve type of coal. The legislature has also funded a flu-gas desulfurization process known as "Sulf-x". This promises to remove sulfur from the gases of a coal plant in the form of elemental sulfur which can be used theoretically for building roads.

Another tremendous oppportunity for Pennsylvania's coal rests in the export market. To that end, we are attempting to assist the private sector in the development of an efficient coal transportation infrastructure. Pennsylvania's major port facility for coal was constructed in 1929. Little more has been done to that facility than routine maintenance. It is in desperate need of modernization, and our allies in Europe and Asia are in desperate need of Pennsylvania's coal. The legislature has just voted a $26 million improvement package for that coal port facility in Philadelphia. Our Department of Transportation is constantly doing road and rail studies to identify which lines need improvement and should be funded for moving our coal out of the United States and to important new markets for steam coal for electrical generation in New England. Our legislature has recently elected to take "primacy" over the regulation of mining activities. This means that the ultimate authority in surface mining would be our Department of Environmental Resources, not the federal governemnt. The Congress of the United States offered this opportunity to each state and we gladly accepted it. Ultimately, an efficient, easy to follow permitting procedure will be established which will allow more mines to be opened in a safe, environmentally sound way. Finally, I would like to sight in more detail the kind of coal development activity which we believe epitomizes the correct role for government in energy development.

Coal mining in the United States began in the northeastern portion of Pennsylvania, in a ten-county area collectively known as the "Anthracite Region." The coal found in that area has special properties which include a very low sulfur content, a high heat content, and excellent storage properties. This coal was the fuel of choice for home heating, industry, and chemical processes from the nineteenth century through World War II. However, with the advent of cheap oil, demand for anthracite coal fell from a high of 100 million tons per year during World War I to about 5 million tons per year in 1980. The result has been massive unemployment and serious economic dislocation for the region. In addition, most mining that occurred in the area predated any environmental regulation whatsoever. Much of the area looks like the target of saturation bombing. However, there's a single project which we have been banking on to re-establish anthracite coal as a major energy resource, create tremendous employment, and reclaim much of the scarred lands without cost to state or federal government. We call this our "Large-scale Project."

In 1976, the Department of Energy funded a Penn State professor named Charles Manula to study the potential for the use of anthracite in electrical generation. Manula concluded that anthracite could be used in a power plant at a cost competitive with other forms of electrical generation if a technique of mining new to the anthracite region was employed and if the Environmental Protection Agency could be persuaded to grant a special exemption to anthracite coal from the scrubbing requirements of the Clean Air Act that recognize anthracite's special low sulfur properties.

Even before Lieutenant Governor Scranton took office, we were approached by the head of one of anthracite coal's associations and asked to develop a state position that would grant to anthracite the necessary exemption. Five days after the Lieutenant Governor was inaugurated, we submitted testimony to the Environmental Protection Agency recommending this exemption. We had four reasons for this recommendation:

1. Anthracite coal burns cleanly, cleaner without scrubbing than eastern bituminous coal is with scrubbing.
2. The old mine scars created in the past along with acid waters and dangerous underground tunnels would be wiped clean by the large, deep open pit technique suggested by Dr. Manula. The private sector would in the course of mining activities pay for an environmental reclamation that the government could never afford. An unprecedented environment cleanup would take place.
3. Significant economic benefit would accrue to that region.
4. An unused energy resource would be tapped.

The EPA agreed — the exemption was granted.

The next step for us was to find utilities in need of additional electrical capacity interested in developing this project. At the direction of the Lieutenant Governor, our Governor's Energy Council went from door to door of utility executives looking for the right one. Finally, the Allegheny Electric Cooperative expressed their interest in the project and was joined by Pennsylvania Power and Light and Philadelphia Electric.

This "Utility Group" argued that federal funding would probably be required to open up the mine necessary for this project. However, we felt that if Manula's studies were right, the private sector was all that was needed. We agreed to find a mining company with sufficient capital to open the mine and sufficient reserves to supply a power plant for fourty years. Several major mining firms have expressed interest in this project, and we believe that all signs indicate a successful initiation of this major project for our northeastern region.

Without expense, without a massive infusion of money from the public sector, without heavy-handed regulation or governmental dicta, a significant project is likely to occur. This role of coordination, assistance, and direction is how government can help fulfill our energy needs to the benefit of all Americans. It is what we consider a good example of how the job can be done with a cooperative spirit between the private and public sectors. We offer to you, Mr. Smith, and to all others, our assistance in pursuing projects like this in energy development.

*Chapter Seven*

# Energy and the Environment

**Robert Leo Smith, Ph.D.**
Professor
Division of Forestry
WEST VIRGINIA UNIVERSITY
Morgantown, West Virginia 26506

Robert Leo Smith is currently Professor of Wildlife Biology in the Division of Forestry at West Virginia University. Before moving to WVU, where he spent the last 17 years, Dr. Smith taught at the State University of New York at Plattsburg.

Dr. Smith earned his doctorate from Cornell University. He has written six books in the field of Environmental Science and Ecology. His book "Ecology and Field Biology" is currently the most popular book in its field in the United States.

The Northeast Section of The Wildlife Society recently honored Robert Smith with its John Pierce Memorial Award for contribution to Wildlife Education. Dr. Smith is considered one of the leading experts on the environment effects of strip mining.

The title of the symposium, *Energy, Economics, and the Environment* suggests a confrontation between economics, growth, and expanding use of energy, and natural resource conservation, between technological development and expansion and the environment.

There are two major views of energy and the environment. One is that expanding energy development and growth equal money, progress, and power. Energy in all forms is a resource to be converted to usable goods. An associated extremist view favors maximum energy development, the environment be hanged.

The other view, often also extremist, is ascribed to so-called environmentalists. This position advocates the banning of nuclear power, reduction of pollution from coal-fired power plants, development of solar power and the use of wood as fuel. Environmental extremists are often regarded by the other side as a hinderance to progress, as people more concerned about snail darters than society.

Both sides are at fault, failing to view the world in a proper perspective. Many environmentalists fail to recognize the energy needs of the country and the world. The economic position fails to recognize that increased energy development can cause severe degradation of the environment. The environment is not the domain of the environmentalist. Rather the environmental degradation affects quality of life including future economic growth.

*Basis of Energy and the Environment*

Energy runs a finite world. That basic fact is usually overlooked by many people concerned with today's energy problem. Energy in the economic and political spheres is viewed as power to run industry and transportation and to provide heat and air conditioning. That is a very narrow view. In reality energy is required to run both the biological and physical world.

With the exception of nuclear power the ultimate source of energy is the sun. Fossil fuels are the excess net production accumulated during the Carboniferous Period. The so-called synthetic fuels are developed from fossil fuels. Wood and other organic biomass are the products of photosynthesis.

Life on earth depends upon a flow of energy. The source of that energy is the sun; its sink is outer space. Solar energy is trapped by plants, converted, and concentrated into a usable form and then utilized until the energy fixed is no longer transferrable.

The basic energy flow on earth is represented by the formula for photosynthesis:

$$\begin{array}{ccc} CO_2 + H_2O + \text{Sunlight} & \longrightarrow & CH_2O + O_2 \\ 397 \text{ Kcal} & \longrightarrow & 114 \text{ Kcal} \end{array}$$

The formula emphasizes a significant fact often overlooked. The dispersed, less concentrated energy of sunlight is concentrated into biomass but at a high energy cost. For each 114 Kcal of energy fixed as glucose, 397 Kcal energy are used. The reaction is possible only because energy from the sun can be paid into the process.

Another point is that as energy is consumed and burned metabolically, energy is released to a more dispersed state:

$$C_6H_{12}O_6 \rightarrow CO_2 + H_2O + \text{energy} \begin{cases} \text{entropy} \\ \text{food chain} \end{cases}$$

The same reaction takes place when the concentrated energy in wood or coal is burned. Potential energy is converted to kinetic energy and released as heat. Thus energy is transformed from a more concentrated to a less concentrated state.

This behavior of energy is described by the second law of thermodynamics. While the first law states that energy can be transformed, can but never be created or destroyed, the second law states that at each transfer of energy, less energy is available to do work. The second law is the entropy law. Energy changes in one direction only; from the usable to the unusable, from order to disorder.

Because energy must be paid into any reaction or process that concentrates energy, the creation of order is done at the expense of creating more disorder in the surrounding environment. This is evident in every activity we undertake. A good example is the production of electricity.

Assume a potential of 100 kilowatt hours of energy in fossil fuels in the ground. Energy costs of extraction reduce that potential by 35 percent or 35 kilowatt hours. An additional 2 kilowatt hours are lost in processing and transportation. That leaves a potential of 63 kilowatt hours at the power station. Of this an additional 39 kilowatt hours are lost in the conversion of coal to electricity. Transmission of electricity over the distribution system requires another 2 kilowatt hours. Six kilowatt hours are lost in the final conversion. That leaves 16 kilowatt hours of useful energy obtained from 100 kilowatt hours of potential energy in fossil fuels.

### *Sources of Energy*

We have available three sources of energy. One is solar energy which is already at work, heating the earth and providing energy for photosynthesis. Out of photosynthesis comes our second form of energy, food and fiber, plants and indirectly animal biomass. This form of energy is called renewable because on a human time scale it can be replaced. The third source of energy is nonrenewable on a human time scale because it cannot be replaced in many lifetimes. These energy sources include coal, oil, gas, lignite, peat, and radioactive rocks.

In spite of this common classification, we fail to consider these sources of energy in the proper context. When we speak of an energy problem, we think of burning fossil fuels, wood, or utilizing nuclear power to keep industry running and home fires burning. We fail to consider food energy which in many parts of the world is also a form of energy crisis. That flow of energy keeps the fires of life going. Life is energy flow. When energy ceases to flow through an organism at the cellular level, the organism dies and entropy sets in.

Unfortunately energy from food and energy for industry and power come into conflict. That results in major energy problems which will become evident later.

Most of our energy problems come about through luxury consumption, encouraged in part by a desire for ever increasing economic growth. Growth requires an increased flow of energy and materials. Material entropy, in the form of solid wastes, industrial effluents, and sewage, is passed on to material or biogeochemical cycles. The input of such wastes results in an overloading of the material cycles resulting in polluted air, polluted water, and polluted land.

An example of luxury consumption can be found in the production of food energy which today is highly dependent upon the input of fossil fuel energy. The situation represents a good example of how increasing an input of energy produces less energy because of entropy costs. Over time we exchanged a low energy flow, low entropy agricultural system for a high energy flow, high entropy system. We exchanged a high energy cost for high individual productivity but low energy efficiency (Hirst, 1974).

For example, a peasant farmer in a more primitive agricultural system produces 10 calories of food energy for each one calorie of human energy expended. A modern Iowa farmer produces 6000 calories of food energy for each calorie of human input (Steinhart and Steinhart 1974). But that low input is subsidized by a high fossil energy input. This subsidy includes pesticides, fertilizer, and fuel for tractors.

The energy subsidy has been increasing over the years. In 1948 yield was 3.7 Kcal per Kcal of fuel input. In 1970 the yield declined to 2.8 Kcal even though corn production increased (Pimentel et al. 1973). By the time one calorie of food energy reaches the American table, 10 calories of energy have been expended getting it there (Steinhart and Steinhart 1974). To this high direct energy cost must be added pollution from leaching of fertilizers and fuel combustion, and soil losses from erosion.

Luxury consumption of material goods results in increased energy costs, increasing burning of fossil fuels, and increased pollution from PCB's, nitrogen, carbon dioxide, sulfur dioxide, and nitroxides in the atmosphere to particulate matter and toxic liquid and solid wastes.

### *Solutions to Energy Problems in Context of the Environment*

The popular solution to the energy problem is to produce more energy and to seek alternative sources. Little emphasis is given to reducing energy flow (energy conservation). In the long run, continued high energy consumption will ultimately result in less available energy at least from conventional sources of fossil fuels.

*Energy from Coal*

One solution to the energy problem boldly emphasized is to turn to coal, primarily because coal in the United States is our most abundant fossil energy resource. In the mid-19th century coal replaced wood as our major fuel (Berg, 1978). Because of the many problems of extracting, handling, burning, and pollution, coal was replaced by oil. Now because of the uncertain supply and high price of oil, the country again is looking at coal.

The problems of coal, however, have not changed and they present some serious obstacles. As the use of coal increases, pollution from burning of coal increases. Although scrubbers, largely restricted to new power plants, remove much of the heavy particulate matter from coal-fired furnaces, sufficient particulate matter and sulfur dioxide escape into the atmosphere to become significant pollutants. For every ton of coal burned, approximately 80 pounds of $SO_2$ is released into the atmosphere. The amount released into the atmosphere in North America is approximately 23 million tons a year. Of this, 19 million tons is produced in the United States, 60 percent of it from the electric power industry.

In the atmosphere sulfur dioxide reacts with water to produce a weak sulfuric acid which comes to earth as acid rainfall. Although sulfur dioxide is not the only source of acid rain, it contributes significantly to it especially over the industrialized regions of the world. Some areas of North America, especially northeastern United States and Canada experience the impact of acid rain (Cronan and Schoenfield, 1979) while other regions such as the Ohio Valley both produce and receive acid rain (Loucks, 1980).

The impact of acid rain depends in part on the nature of the soils and geology of the region receiving it. In areas of highly buffered soils, acid is neutralized in the soil and has had little affect so far on the terrestial and aquatic ecosystem (Dochinger and Seliga 1976). Evidence from northern Europe indicates that acid rain can reduce forest growth (Braekhe, 1976). Another effect of acid rain not highly publicized is the financial loss incurred from corrosion of metal in industrial buildings and cars and the deterioration of stonework made from marble and limestone. Although the problem of acid rain is not well understood, a serious acid rain problem does exist in spite of frantic efforts of coal and electric power industry to downplay the problem. But in regions with acidic soils and sandstone and granitic geological structures, the situation is different. Weakly buffered streams and lakes can be turned into highly acidic ones, a condition that eliminates aquatic life, particularly salmonid fishes.

An associated problem of burning coal is the increased amount of carbon dioxide added to the atmosphere. The amount of $CO_2$ in the atmospere has increased 18 percent in the past 100 years from 260 to 300 ppm (parts per million) to 330 ppm (Siegenthaler and Oeschger, 1979). This increase accounts for 50 percent of the total amount of $CO_2$ added over that period of time. The other 50 percent apparently has been scrubbed from the atmosphere by oceans, shallow water sediments, and terrestrial vegetation. By the year 2000 the $CO_2$ theoretically could rise to 385 ppm.

The significance of this increase in $CO_2$ is its potential for warming up Earth. Carbon dioxide in the atmosphere tends to prevent the escape of outgoing long wave radiation from Earth to outer space, the energy sink. As more heat is produced and less heat escapes, the temperature of Earth increases. For every 10 percent increase in $CO_2$ in the atmosphere, the average air temperature of Earth increases 0.32 °C. A doubling of $CO_2$ in the atmosphere would result in a 2.4 °C rise in Earth's temperature (McLean, 1978).

A warming of the atmosphere would have a profound effect of Earth. Warming of polar waters would affect the density stratification of the oceans. As the oceans warmed more $CO_2$ would be released from the oceans to the atmosphere. A 1 °C rise in the surface temperature of the ocean in the regions of deep water would increase atmospheric $CO_2$ by 4 percent. The oceans hold 60 times as much $CO_2$ as the atmosphere. If 5 percent of deep oceanic waters released their $CO_2$, atmospheric $CO_2$ would increase by 25 percent. A rise in Earth's temperature could damage animal reproductive systems (McLean, 1979), accelerate the melting of ice caps, raise the level of the oceans and change the climate regionally and globally. Although any predicted increase in Earth's temperature from increasing $CO_2$ is not detectable to date, the effects may be delayed or masked because of compensating cooling from increased particulate matter and aerosols in the atmosphere and perhaps thermal inertia in the oceans (Madden and Ramanathan, 1980). The natural cooling trend of the past decade may reach a minimum in the next decade and the onset of $CO_2$ induced warming could take place rather suddenly (Broecker, 1975). The point is that the potential for increasing the temperature of Earth from accelerated burning of fossil fuels is very real and must be considered in any long range energy planning (Broecker et al, 1979). The trend once started could be irreversible.

Another aspect of burning coal is the release of radioactive substances to the atmosphere from coal fired power plants (McBride et al., 1978). If one assumes a 1 percent ash release to the atmosphere (EPA regulation) and 1 ppm of uranium and 2 ppm of thorium in the coal (both a U.S. average) then the amount of radioactivity released from coal fired power plants (1000 megawatts) is as high or higher than radiation released from pressurized water or boiling water nuclear reactors of the same size. Because the average ash release from power plants is much higher, up to 8 percent, and coals having higher concentrations of uranium and thorium are common, the amount of radioactivity released from coal fired power plants could be a magnitude greater. The public health effects of these levels are minor, much less than the airborne releases of nonradioactive material such as particulates and $SO_2$, but the comparison with nuclear power plants is interesting.

*Surface Mining.* Surface mining provides over one-half of the coal produced in the United States and will continue to do so in the future. Although reclamation practices, when enforced, can reduce erosion and restore some cover and stability to the disturbed land, reclaimed land is not suitable for highly productive agriculture. No one has yet demonstrated that prime agricultural lands, once disturbed by surface mining, can be resored to their original productivity. Even advanced reclamation practices cannot replace a native forest with its rich diversity of species, rich herbaceous understory, its vertical structure and wildlife. Reclamation in semi-arid western rangeland and wetlands is difficult because of a lack of water and an understanding of how semi-arid ecosystems function. Successful revegetation in that water-scarce region depends upon irrigation and water is allocated to other uses. Further, we have no knowledge of the ability of restored rangelands to withstand intensive grazing.

The location of the major coal fields of the United States emphasizes the conflict between coal, food production, and timber production (Fig. 1). Much of our future coal production will come from the cornlands of the midwest, the wheatfields of north central United States, the rangelands and irrigated croplands of the west, and the most productive hardwood forest regions in the world. By overturning these lands for coal, we will sacrifice one form of energy, renewable food and fiber, for another: nonrenewable fossil fuels. Such sacrifices should not be made without serious study of the potential long range impact on our future welfare.

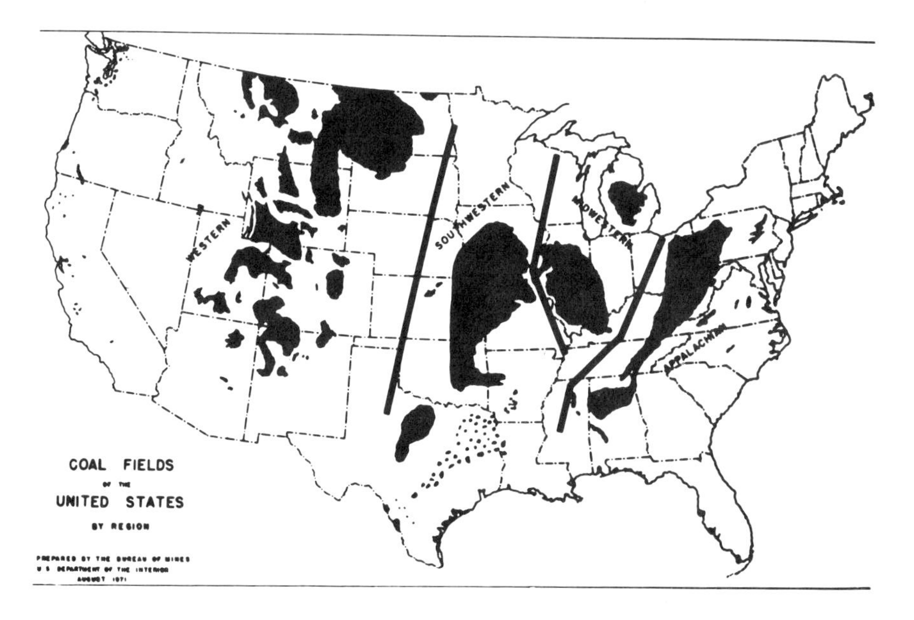


FIGURE 1. A map of the coal fields of the United States. Note the location of the coal fields relative to agricultural regions.

*Energy from Biomass*

"Split wood, not atoms" is a popular slogan of those who argue for the use of biomass, particularly wood and crops, as a so-called renewable energy source. The assumtion is that wood and other biomass burned can be balanced by new growth. However sound such a solution may appear on paper, it has a number of shortcomings. That is apparent when you compare current energy needs with the potential of renewable biomass. Annual energy consumption in the United States is $18 \times 10^{15}$ Kcal. That amount is greater than total solar energy fixed by plant biomass in the United States, which amounts to $13.5 \times 10^{15}$ Kcal annually. Of this, 42 percent is produced by forest and 6 percent by cropland. Nearly one-half of solar energy fixed is achieved by agricultural cropland and forests (Pimentel et al., 1978).

If one assumes an average of 5 ton of biomass produced per acre, an equivalent of 6000 Btu, 1.27 billion acres or 1.97 million square miles of land would be needed to supply all our fuel needs. Such a requirement would leave little for food production.

The total solar energy conversion by two major biomass producers, forest and cropland, amounts to $5.8 \times 10^{15}$ Kcal which represents about 32 percent of fossil energy conversion. The total energy conversion by other biomass sources could produce $0.34 \times 10^{15}$ Kcal or 1.9 percent of current fossil energy use (Pimentel et al., 1978).

A second problem with biomass conversion is the conflict with another important energy requirement: food and fiber. Seventy percent of the land area of the United States, and one half of the solar energy fixed by plants is used for the production of food, fiber, and forest products. Of this 66 percent of harvested energy is in pasture and forage crops, 18 percent in forest products, and 16 percent in food crops.

A strong suggestion widely publicized is to use manures, crops, and lumbering residues. While they do represent potential energy, their use could have an adverse effect of future energy fixation by plants. For example, crop residues from grain corn production amounts to 6.7 ton per hectare. To remove this residue from fields would reduce organic matter in the soil, adversely affect soil structure, decrease its water holding capacity, increase soil erosion, and reduce nutrients in the soil. Corn residues amounting to 123 kg/ha in an Iowa corn field contains 40 percent of the nitrogen, 10 percent of the phosphorus, and 80 percent of the potassium of current fertilizer application. To replace that loss with commercial fertilizer would mean the expenditure of 2 million Kcal/ha/yr. or 200 liters of petroleum. Energy wise, more would be lost than gained (Pimentel et al., 1978).

As for residue in lumbering, 7 to 20 percent of the biomass is left in the woods. In some operations this waste is used in the production of synthetic board and is too important a resource to burn. While leaving some slash behind in the woods after harvesting may appear wasteful, logging debris is an important source of nutrients to forest soils. Nutrient balance in the forest depends partly on short term annual leaf fall and longer term cycling through twig and wood decomposition.

Losses of nutrients through harvesting can be considerable. For example tree length harvesting of a spruce-balsam fir stand removes 150 kg per hectare of calcium, 79 kg of nitrogen, 27 kg of potassium, and 52 kg of phosphorus. Whole tree harvesting results in the per hectare loss of 413 kg of calcium, 387 kg of nitrogen, 159 kg of potassium, and 52 kg of phosphorus. This contrasts with the 117 kg of exchangable calcium in soil, 14 kg of exchangable nitrogen, 65 kg of exchangable potassium, and 6 kg of phosphorus. Removal of forest debris and whole tree harvesting result in nutrient depletion, reduced forest growth, and changes in forest composition. These impacts do not take into account the effects of wood harvest on wildlife, soil structure, soil erosion, recreation, and water quality.

In spite of the arguments of proponents of wood as fuel, burning wood is polluting (Allaby and Lovelock, 1980). While the smell of wood smoke may be much more pleasing than the sulfurous smell of coal smoke, wood is rich in polycyclic organic chemicals, all well-known or suspected carcinogens. Like burning coal, wood burning also produces a great deal of atmospheric particulate matter.

Biomass proponents to the contrary, the only realistic biomass sources ecologically and economically available are food processing, urban and industrial wastes, and sewage. They could provide about $64 \times 10^{12}$ Kcal of energy, about 0.36 percent of annual U.S. energy and consumption (Pimentel et al., 1978).

Also strongly suggested is the production of alcohol from grain particularly corn and wheat. Aside from the sheer folly of converting food energy to industrial energy, there are strong economic and entropy arguments against it (Weisz and Marshall, 1979). About 1.3 bushel of corn will produce 1 gallon of fuel, but the production of a gallon of grain alcohol requires 2 to 3 gallons of high grade fuel. Thus is takes more oil to produce corn and ferment it into alcohol than is saved by using alcohol as fuel.

Based on an average production of 90 bushel per acre, one acre of corn would produce 55 gallons of fuel. If the entire corn production of the United States, approximately 75 million acres were devoted to fuel production, it would provide only 3.7 percent of the 1977 U.S. gasoline requirements and 1 percent of U.S. petroleum production. Not one bushel would be left for human consumption. The absurdity of using grain crops for fuel is obvious.

*Synthetic Fuels*

A well-publicized potential solution to the energy problem is synthetic fuels produced from coke, especially solvent refined coal (SRC).

Solvent refined coal production can produce some enormous environmental problems. Putting aside all the dangers of environmental contamination with powerful carcinogens and toxic solid wastes, one is still left with massive water allocation problems involving withdrawal, consumption, allocation, and quality. Hydrologists point out that even in the humid Eastern United States water allocation will be a major environmental and economic problem in the coming decade of nearly the same magnitude as that of oil.

SRC plants compound the problem of water allocation because they are such heavy consumers of water, utilizing most of the water they withdraw from rivers and lakes. Withdrawal is the amount of water drawn from a source; consumption is the amount of water used and not discharged back to the source. As an example, an average coal gas conversion plant may withdraw $10^6$ $m^3$ of water annually from a river (Hale and Gusseir, 1978). Of this, $10^5$ $m^3$ per year is used in the conversion process. Of the amount withdrawn, one-third ($3 \times 10^5$ $m^3$) is lost to evaporation. Evaporative losses are not returned to the immediate area but fall as rain somewhere else. Of the remaining amount, $1.5 \times 10^5$ $m^3$ remains as heavily polluted water sent to a treatment facility; of this $10^5$ $m^3$ is evaporated, 50,000 $m^3$ is lost as waste products and 450,000 $m^3$ is returned as treated water to the source. Of the 1,000,000 $m^3$ withdrawn, 550,000 $m^3$ is consumed. In effect, 55 percent of the water withdrawn is unavailable to downstream users.

In comparison to other users that is high consumption. Industrial and mining and manufacturing consume 11 percent of withdrawal, and thermal power plant cooling 1.4 percent. Rivaling coal conversion is irrigation with 57 percent consumption.

As an example of the potential of increased production of synthetic fuels to divert water from other uses, one can consider the proposed Solvent Refined Coal Plant (SRC-II) to be located on the Monongahela River at Morgantown, West Virginia. This demonstration plant, if built, will be underwritten by the governments of the United States, Japan, and Germany, and a private company, SRC International. The plant would have a capacity of 6000 tons per stream day (tpsd) of fuel oil. If successful, the plant would be expanded to a commercial operation of 30,000 tpsd (Department of Energy, 1981).

The proposed SRC-II plant would consume a greater percentage of withdrawal water than a coal gas conversion plant. The demonstration plant would withdraw an estimated 304 liter/second (1/s) (10.7 cfs) from the Monongahela River, sidcharge 78 liter/sec. (2.7 cfs) and consume 226 1/s (28 cfs) or 78 percent of the withdrawal water. A commercial plant would consume 793 liter/sec. (28 cfs).

The Environmental Impact Statement argues that the SRC-II demonstration plant would consume only 0.2 percent of Monongahela's mean flow; but during low periods it would consume 2.4 percent of the river's flow at the plant site. A commercial plant would consume 10 percent of low flow at the plant site.

Such a plant, both demonstration and commercial, would have a pronounced effect on downstream water users. During periods of low flow, consumptive use would be sufficient to interrupt barge traffic and force municipalities and industries to curtail withdrawal. Because in Pennsylvania navigation receives priority, water allocation would go to navigation over industrial users.

In addition to a shortage of available water, water quality in the river would also deteriorate. At present, water quality during low flow in the Monongahela River is not able to meet Pennsylvania quality standards for total dissolved solids (TDS), pH, and dissolved oxygen (DO). The SRC-II plant would further aggravate conditions. Added to the current pollutants, which in past decades have been declining, would be an increased load of phenolic compounds, toxic elements, and salt. Plans call for the discharge of 22 tons of salt a day from the demonstration plant into the Monongahela River and 120 tons of salt per day from the commercial plant. The EIS indicates that the river would be able to assimilate this discharge 60 percent of the time. The addition of salt, a highly corrosive substance, to a fresh water system, however, would upset ecological conditions and could create severe problems for water consuming industries and municipalities downstream to Pittsburgh.

*Oil Shale*

Oil shale found in northwestern Colorado, eastern Utah, and southwestern Wyoming has the potential for providing enormous amounts of oil (Metz, 1974). Exploiting this resource requires the extraction of oil from the shales. Such exploitation will result in large scale destruction of western grazing and wildlife lands, production of enormous amounts of spent rock, destruction of underground aquifers in the region, and an intolerable demand on western water (Metz 1974, Gardner and Lebaron 1968). Water use for shale oil will conflict with water for surface mining of coal. Water for irrigation and domestic use would have to be sacrificed with adverse effects on food production. In addition, successful revegetation of spent shales is questionable, in part because of the lack of water to establish and maintain the plant cover; and salt-laden runoff from bare shale has the potential of polluting the Colorado River.

*Nuclear Power*

Nuclear power supplies about 4 percent of annual energy supply although in some regions as New England it may supply up to 60 percent. The problems with nuclear energy are so well known that they do not need to be elaborated here. The problems at Three Mile Island and at Japanese nuclear power plants underscore the situation. In spite of these incidents, nuclear power in many ways is environmentally cleaner than coal. The major problem with nuclear power is the disposal of nuclear wastes including the dismantlement of old power plants. We still have no well-developed plans or site locations for long term storage of highly radioactive wastes that will remain dangerous for thousands of years. Such wastes are now held temporarily under water in specially constructed pool and storage tanks.

### *Solar Power*

Solar power is the energy solution of many environmentalists. Solar power has its greatest potential in smaller structures, such as houses, but it offers no major solution to the energy problem. What is overlooked by solar power advocates is the information represented in the photosynthetic equation. Solar radiation represents the entropy of the sun. Concentration of this dispersed energy requires a large input of energy to build and install the necessary equipment (Whipple, 1980). A rapid conversion to solar power, according to best estimates, would involve an investment 1 to 1½ times the value of energy gained. In addition to cost, there are certain environmental concerns. Solar cells needed to concentrate sufficient sunlight to provide current needs would cover 10 to 20 percent of the United States land mass. Another suggested means of obtaining solar energy is to place giant solar collectors in space to beam back concentrated solar radiation to earth. Microwave radiation from beaming back energy would make areas about the collectors uninhabitable.

### *Water Power*

Water power provides about 3.5 percent of our annual energy needs, which is about the maximum possible. A small additional amount could be produced by adding hydroelectric systems to existing flood control dams. The opportunity for building new dams is limited because of high costs, the limited number of sites remaining, and the need to maintain the integrity of our rivers. Further dam construction would eliminate remaining prime bottomland farms needed for production of food. Pumped storage, pushed by power companies as a source of peak-use power, is energetically inefficient. More power is used to pump water up to the high level storage dams than is produced by the release of the water through the turbines. The main argument for pumped storage is that power produced during low load periods could be used to pump water up for use during peak load periods, but production involves a net energy loss. Energy conservation is more effective.

### *An Approach Toward a Solution*

Are there solutions to the energy problem that will still allow a viable environment? The answer is "yes," but the solution can be painful to a society brought up on luxury consumption.

The short term solution is to make best use of various sources of energy, without relying solely on any one. The long term solution is a reduction in energy flow-through, in effect a reduction in entropy. That amounts to energy conservation. Although energy conservation can significantly reduce energy demands, it is not popular with the power industry or government, primarily because a reduction in energy consumption is equated with lower economic growth.

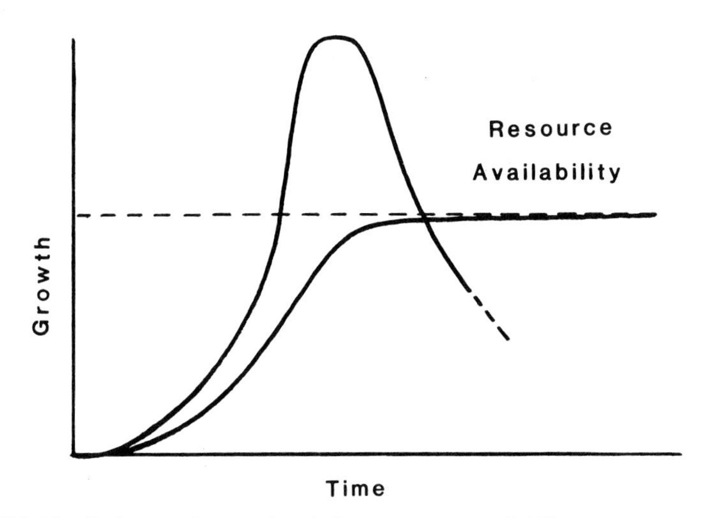


FIGURE 2. Two basic growth curves in relation to resource availability. The one curve represents exponential growth of a population that exceeds resource limitations and experiences a sharp decline. Such curves are typical of populations dependent upon a non-renewable resource. The second curve is sigmoidal. Population growth or economic growth declines and levels off as the limits of resources are reached.

Increased energy consumption and economic growth cannot continue indefinitely. Perpetual growth is contrary to all physical, biological and ecological laws. It is contrary to the second law of thermodynamics. The growth of organisms and the growth of populations are limited. Organisms grow until most of the energy intake is used for maintenance and tissue replacement. Growth of populations is limited by resource availability (Fig. 2). Early in its history, the population may experience luxury consumption, but as its density increases, the amount of resources available for each individual declines. Population growth then declines and levels off as it reaches the limits of available resources. If a population's growth is continuous, then that population ultimately experiences a resource depletion and declines suddenly and precipitously.

The world is finite for human populations, too. There are limits to resource availability and thus limits to economic and energy growth. Declines in economic growth in the past were halted and growth then stimulated by the availability of new resources. In the Middle Ages the Black Death removed over one-half of the European population, reducing competition for wealth and land. The economic problems and overpopulation in Europe during the Industrial Revolution were corrected by the discovery and subsequent colonization of a resource-abundant New World. Now the world has run out of land, resources are diminished, and the population is still growing. We are approaching the limits of growth whether we want to admit it or not.

This idea does not rest well with traditional economists (Miernyk, 1981) whose basic premise appears to be perpetual growth in a finite world. Such growth ob-

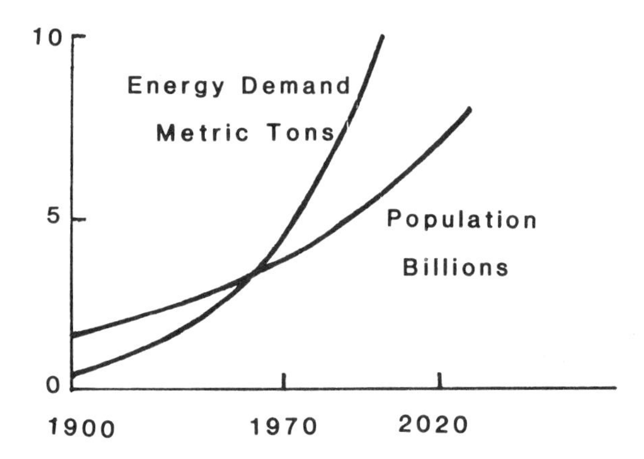


FIGURE 3. Relationship between world population growth and energy demand in terms of billion metric tons of coal equivalents. Energy demands are exceeding population growth which still continues. When one considers that energy resources are finite, the relationship of these graphs to Figure 2 becomes obvious. (Adapted from B.F. Fogel, "Thinking About Energy" *RF Illustrated,* April 1980, Rockefeller Foundation.)

viously is impossible, but most economists find a declining or zero growth untenable. While they will accept that on a regional scale, they will not accept low growth on a national scale.

Unlimited growth ultimately will result in a crash from which it will be difficult to recover in the face of depleted resources. Both the world population and energy demands have been growing exponentially since the middle of the last century. Before 1950, however, energy demand was well below population growth (Fig. 3). During the 1960's energy demands increased faster than the population and the two growth curves crossed. During those years, energy supplies kept up with demands, but in the 1960's energy supplies grew more slowly than demand. Increased competition for limited energy resulted in increased prices and periodic shortages, constraining economic growth. The task now facing the world, especially the major energy consumer, the United States, is to bring the population growth and the energy demand curve into equilibrium.

Some will look for new unlimited quantities of renewable energy that will continue to support unlimited growth. But unlimited growth is contrary to the second law of thermodynamics which holds for a world of growing populations and declining energy supplies. Once we have recognized that the second law will not be repealed, we can begin long-term energy planning, including production of food and fiber, and the maintenance of a livable environment.

## LITERATURE CITED

Allaby, M. and J. Lovelock 1980 Wood stoves: the trendy pollutant. *New Scientist* (13 Nov. 1980): 420-422.

Berg, C.A. 1978 Process innovation and change in industrial energy use. *Science* 199:608-614.

Braekhe, F.S. (ed) 1976 *Impact of Acid Precipitation on Forest and Freshwater Ecosystems in Norway.* SNSF Project NISK 432 Aas-NHL, Norway.

Broecker, W.S. 1975 Climatic change: are we on the brink of a pronounced global warming? *Science* 189:460-463

Broecker, W.S. et al. 1979 Fate of fossil fuel carbon dioxide and the global carbon budget. *Science* 206:409-418

Cronan, C.S. and C.L. Schofield 1979. Aluminum leaching response to acid precipitation in high-elevation watersheds in the Northeast. *Science* 204:304-306.

Department of Energy 1981. Final Environmental Impact Statement-Solvent Refined SRC-II Demonstration Project. 2 Vol. Department of Energy, Washington, D.C.

Dochinger, L.S. and T.A. Seliga (eds) 1976. *Proceedings First International Symposium on Acid Precipitation and the Forest Environment.* U.S. Department of Agriculture, Washington, D.C.

Gardner, B.D. and A.D. Lebaron 1968. Some neighborhood effect of oil shale development. Natural Resource Jour. 8:569-585.

Hale, J. and M. El Gasseir 1978. Energy and water. *Science* 199:623-634.

Hirst, E. 1974. Food related energy requirements. *Science* 184:134-138.

Loucks, O.L. 1980. Acid rain: living resources implications and management needs. *Trans. North Amer. Wildl. and Nat. Resource Conf.* 45:25-37.

Madden, R.A. and V. Ramanathan 1980. Detecting climate change due to increasing carbon dioxide *Science* 209:763-768.

McBride, J.P. et al. 1978. Radiological impact of airbourne effluents of coal and nuclear plants. *Science* 202:1045-1050.

McLean, D.M. 1978. A terminal Mesozoic "greenhouse": lessons from the past. *Science* 201:401-406.

Metz, W.D. 1974. Oil shale: a huge resource of low grade fuel. *Science* 184: 1271-1272.

Miernyk, W.H. 1981. The prospect of perpetual decline. *West Virginia Alumni Magazine* 4(1): 6-8.

Pimentel, D. et al. 1973. Food production and the energy crisis. *Science* 182: 460-463.

Pimentel, D. et al. 1978. Biological solar energy conversion and U.S. energy policy. *Bioscience* 28:376-380.

Siegenthaler, V. and H. Oeschger 1978. Predicting future atmospheric carbon dioxide levels. *Science* 199:388-395.

Steinhart, J.S. and C.E. Steinhart 1974. Energy use in the U.S. food system. *Science* 184:307-316.

Weisz, P.B. and J.F. Marshall 1979. High-grade fuels from biomass farming: potentials and constraints. *Science* 206:24-29.

Whipple, C. 1980. Energy inputs of solar heating. *Science* 208:262-266.

*Panel Discussion*

# Energy, Economy and the Environment

## Panel Discussion

*Justice Flaherty* — We are now to the part of the program which calls for questions to the participants of the symposium. We hope that the views which are expressed will be used for reference material, not only in scientific quarters, but in legislative and judicial quarters, as we deal with the subject of the energy and the environment. In my own instance being justice of the highest court of the third largest state in the United States, I am particularly aware of how important the judiciary is in this general subject. It is to us in the final analysis that all of these matters eventually get; therefore, it is very important that the viewpoints be well known. The judicial, legislative, and executive branches of the government are all extremely interested in this area. As a non-professional scientist, I have been greatly impressed with the problems that are facing us. I have numerous questions that have been submitted.

This question is addressed to *John Taylor* — "James Smith mentioned that labor was not convinced nuclear energy was currently being used safely. What do you consider the main problems in nuclear energy safety, or are there?"

*Mr. Taylor* — As I mentioned in my talk, there is no method of reducing energy that does not have some safety impact and, therefore, some element of risk in it. We must take some risks in whatever we do, even crossing a street as we leave this building. Where are the safety issues in nuclear power? During normal operation of that plant, controls have been established to collect the normal emissions of radiation within the system so that there is essentially no danger to the public as that plant is normally operated. The levels of emission from the plants, in fact, have been reduced by a factor of 100 over the past 20 years as more knowledge has been gained about potential effects of low-level radiation. There is less radiation totally, in fact, emitted from a nuclear plant than from the natural radioactive elements in the combustion products that come out of the stack in a coal plant. So, the issue is "will an accident occur which will create hazards to people?"—the type of accident that did occur, in fact, in Three Mile Island. To protect against injuring people in the event of an accident, major precautionary steps, containment measures we call them, are built into these nuclear power plants. There are literally five levels of containment in those systems. The fuel within which the high-level radiation exists is first encapsulated in a ceramic material which is relatively impervious to corrosion. Those ceramic fuel pellets are inserted within metal tubing which is very carefully manufactured and inspected so that the radiation within those pellets, in turn, is trapped within that tubing. Those tubes are, in turn, put into a very thick steel pressure vessel which furnishes the third barrier. The pressure vessel is installed in a concrete cell which provides a fourth barrier. Finally, the fifth barrier is the oval-shaped containment vessel—not the cooling tower—not that elliptical-shaped tower, which most of the television programs show you when they show you Three Mile Island, that's the cooling water system—but a much smaller building that's almost inconsequential in size by comparison to the cooling tower, and that's the fifth barrier. Now the Three Mile Island accident was a big one, but the fifth barrier had been put there in the event it happened, and it performed its function. The containment building held back the radiation, and no one in the vicinity of Middletown was hurt by the small amount of radiation that was emitted.

We have much more work to do so that we don't even have that kind of an accident, because the economic damage caused by that alone is extremely serious. We are training the people who run those systems better, providing control rooms that are easier to operate, and so on.

Now, there are other parts of the system which involve radiation such as handling the fuel and sending the waste products from the plant to be stored, and so the next problem in safety is the handling of fuel shipments and the disposal of waste products. We believe that we have the technical resources to do that very safely. In the military program, these things have been done for about 25 years with standards that were developed in wartime emergencies and have been very slow to be changed; those standards are much, much less rigorous than those by which we are governed. And yet no one, in spite of some accidents, has been hurt in the process of transporting and storing those waste products. Our biggest task is to prove by doing, by demonstrating, that storage of those wastes is safe. Laboratory tests have all shown that it is safe. But it has been essentially impossible politically to implement a demonstration program for storage of nuclear plant wastes. One big nuclear power plant like the one we have down at Beaver Valley will generate waste in a year that can be stored in 10 canisters about a foot in diameter and 10 feet high. But, we are being prevented by political opposition and regulatory confusion from demonstrating that it can be done safely.

*Justice Flaherty* — This question was submitted for *Mr. James W. Smith* — "Do you believe that the Federal Government is a better vehicle for developing alternate energy sources to the point of commercialization that is private industry? If so, how do you evaluate the performance of Conrail and the Post Office?"

*Mr. Smith* — I think that the public has to find a mechanism for providing industry with the incentives and the rewards that are needed to make some transition in our energy systems. I hold no great faith in the Federal Government of the United States as currently operated or as it's likely to be operated in the next four years. I have my problems with bureaucracies there and elsewhere. We are afflicted with bureaucracies whether we are universities, trade unions, corporations, or government agencies, and none of them work as well as we'd like to have them work.

The Post Office work was done a "hell" of a lot better when it was the property of the Federal Government than when it was set up as an independent corporation, though I will have to say that I don't think these things can be judged rhetorically. I think we have to make practical judgments.

I was very much concerned about the prediction by Professor Smith, I believe it was, who showed us a picture of the curve of increased production leveling off and who informed us that we had to prepare ourselves for a society of no further growth.

If I believed that, I would say to you without any hesitation, we must prepare ourselves for a society operating under some other form of economic system than capitalism. Capitalism, our whole system of private ownership of business firms, is predicated on growth and won't function without. If you don't understand that, go take a course in sophomore economics from somebody that will teach you macro- and micro-economics and you'll understand it. That's the way it is. If we have to plan for no growth, then we have to plan for public operation and public planning of our productive facilities.

I don't think that we do have to make that decision. I very much hope we don't, but I do think that to make some of the conversion we now need to make that we have to have public intervention in the economic system.

*Justice Flaherty* — Another question for *Mr. Taylor* — "What is the current status of technology related to disposal of radioactive wastes?"

*Mr. Taylor* — I stated earlier and broadly about that ability, but let me answer the second question more specifically. Other countries are taking active measures to store the wastes from their nuclear power plants. The French have a waste disposal facility which uses the modern methods which we have developed in the laboratory. The Germans are in the process of building a facility. The Swedes, you recall, had a very, very major confrontation on nuclear power, which was favorably decided in a referendum just this year. Before the referendum was even held, it was required that the government establish a method of nuclear waste disposal from their power plant. They did define it; their scientists reviewed it; they decided it was adequate; and they are proceeding to implement it. So, around other parts of the world, this waste problem is being effectively handled. I make one comment on Professor Smith's remark that they are dumping these wastes in concrete form into the ocean. That was done years ago for low-level wastes, wastes that are not as dangerous or more typical of the radiation you see in uranium mines. But it has been stopped just on a general precautionary basis, no assurance that it's really hazardous. High-level waste has not been dumped and was never intended to be dumped into the ocean.

*Justice Flaherty* — This question is addressed to *Mr. Goldhaber* — "In terms of recycling, what, if any, statewide programs are active? If there are none, what can be done to initiate discussion and, or eventual action in passing a state recycling law?"

*Mr. Goldhaber* — Let me answer it two ways. If you mean specifically laws mandating the recycling of bottles and cans, I will be candid with you; I don't think that the state legislature is in a mood to do something like that. There were several bills proposed during the previous administration, and they were never allowed to get out of the committee. There is serious deep-seated opposition to that kind of effort.

With respect to recycling of waste oil, there is a project that the energy council has been sponsoring; admittedly, it is largely cosmetic but it's a step. It is not many gallons of oil, but there are recycling centers where you can drop off waste oil, if you change your own oil in the car. There are a number of interesting projects going on in recycling of paper wastes and other biomass products, using them for energy generation rather than land-fill. I'm sure you are all familiar with the kind of paperwork that comes out of Harrisburg. Well, we burn it and it heats the whole capitol complex. True, mixed with a little added garbage from the city. There are fascinating projects using the recycling of manure in agricultural applications for electrical generation with methane. It's going on at a farming community in Gettsburgh; and of course, there are all kinds of things going on at Rodale Farms for experiments in the value of recycling biomass and animal wastes. So, there is something going on, and we certainly encourage it. But as far as bottles and cans are concerned, may I throw the challenge back to you; make your legislator aware of the need to do this kind of thing, and mount a lobbying effort. Then there is a good chance. Let me add that economics is always the most persuasive argument for recycling, and that you must be armed with sound data before you begin the attack. Far more important is encouraging voluntary recycling.

*Justice Flaherty* — If there are those who would propose a more affirmative approach to recycling in Pennsylvania, certainly those people would be advised to use the legislative process, and to organize and make known their views to the legislators. Here's a question directed to *Dr. Robert L. Smith* — "Doctor, would you comment on the contribution of nature to pollution problems; particularly, what is the effect of Mt. St. Helens, plant respiration, decaying animal and vegetable matter, volatile vegetable products?"

*Dr. Smith* — That's a question that's thrown around a lot. I don't worry too much about pollution from nature. The amount of pollution you are going to get from the volcanic eruptions, organic matters, decompositive and the like, is very small relative to sulfur dioxide and carbon dioxide released to the environment daily by industrial society. We don't need to worry about the vegetation from the Smokey Mountains giving off some volatile materials, which the trees do, but these are local—of no environmental concern, for without this vegetational haze the mountains would not be the Great Smokies. In addition, the material is not toxic. I think you've all heard about the Mt. St. Helens statement made by President-elect Reagan. He put his foot in his mouth on that one—I heard that exact figure from the head of EPA when he responded to Reagan's statement concerning the amount of sulfur put into the air by Mt. St. Helens compared to all the car production in the United States for one day, and I cannot recall it. Of course, the amount of sulfur dioxide from the volcano was considerably less. The main impact of Mt. St. Helens might be more the release of particular material in the atmosphere which might change temperatures temporarily, but I wouldn't worry about that either. There is much more particulate matter being poured into the air per day by power plants in the United States than Mt. St. Helens contributed. There is an exact figure on that, and if Ms. Nicholson from the EPA were here this afternoon, she would probably answer that for you, because apparently her boss gave the correct answer to Reagan's charges.

*Justice Flaherty* — This question is addressed again to *Mr. Taylor* — "You stated that the way we should go is to change to electricity as much as we can. Hydrogen produced from electrolysis of $H_2O$ by solar or wind and shipped via pipeline to market is also, it seems to me, very attractive, especially, since non-removable fuel is not used as would happen in making electricity only. Why does hydrogen receive so little attention?"

*Mr. Taylor* — That's a good question. We at Westinghouse have been working very hard on the issue of producing hydrogen by electro-hydrolysis methods and others, and the big problem is the energy required—the efficiency is so low, that the energy requirements to generate the hydrogen are still well beyond our technology. We've looked at every system we can think of, but we can't come up with a sufficiently efficient method of generating that hydrogen to make it economic. If we could it would be a marvelous additional method of serving our energy needs. But it seems still out there on the horizon. It is still well worth working toward, as all these things are, but we see it way out in time, yet.

*Justice Flaherty* — Here is the final question of the symposium — "Since we have seen today the diversity of solutions to the energy problem, and with substantiating evidence, do you as a group foresee a way to evaluate this evidence and compromise a common solution?" That's certainly a profound question, and I'm certain that there can be no immediate answer. I can add this before I submit it to the panel, that the intent of this seminar and this symposium was to provide a broad-based exposition of viewpoints, done in a very responsible atmosphere, and we think we've accomplished that. Each panel member will be given this question.

*Mr. Goldhaber* — It's difficult to know how to analyze the directions to take, the complexities of the market place. I think today was a typical example. Mr. Smith and I have had a little disagreement that shows that there are different ways of looking at the same problem and coming up with divergent conclusions. However, what is needed is a consistent, unified direction guided by a certain set of principles and then see whether or not the situation improves as a result of following that path. I trust that we will see something of that nature in the next 4 or 8 years. I think there is reasonable concensus among ourselves on what paths to take, and we'll see if it works. If it doesn't, we'll be happy to step down.

*James W. Smith* — I would say that it certainly can be done. There are societies that are solving these problems much more efficiently that ours. I would say that Northern Europe generally is, and I think in Japan they are, under a diversity of political structures. In Northern Europe, essentially the labor movement in most of the countries involved has provided the governments for many years; in the case of Sweden from the 1920's until 1978 or '79, in Germany for the last 7 or 8 years. At the other extreme, the business system of Japan provides the government ever since World War II, since they've had a democratic government. So, I don't think whether or not we achieve compromises that make it possible for us to begin to solve these problems rationally and constructively and efficiently depends necessarily on which elements of society are in the political ascendancy or supremacy at the time. It can happen from either direction. What it's going to depend upon is a willingness on the part of all of those involved to recognize that there are common interests in the political system itself—in the survival of the culture—which perhaps are superior to the particular interests of each group in society. If we can achieve that, then I think we can compromise the differences; we can achieve solutions effectively. I believe our technologies, our scientists, our engineers, are capable of doing this. Our problems are much more political than they are scientific.

*John Taylor* — I think I expressed my conviction that we must continue to have economic growth, and it must be fueled by energy growth. This country's requirements are tiny compared to those countries such as India where people are still in such a substandard state. The big issue then is, can we do it? I think that's a matter of faith. When I started high school, no one knew there was such a thing as fission, no one knew it was possible to get energy out of the earth or out of water, which is going to be the case with fusion. It wasn't even conceived. I believe that if we take out technological effort and mesh it with the legitimate concerns of the environment and our social needs, we will achieve that growth and we will have put behind us the problems we seem to be wallowing in so much today.

*Justice Flaherty* — As far as my comment, I should make note that I represent something which has not been addressed in this symposium. The symposium has focused upon energy; energy as a physical thing. However, I represent the law which is really the energy of the living world, and the law is developed and, in our Anglo-American society, is defined by our judiciary. In the final analysis, the judiciary must accommodate the various solutions which will be forthcoming. I hope that my brothers have the foresight and the stamina to accommodate what might be quite novel innovations in the law, which is the living energy, to make this world a place in which it's worth living, since that is the function of the law.

*George Leney* — The question, as I understand it, is can we solve our energy problem? My answer would be yes; we are doing it. The critical question is not whether we can do it or not, it's the time frame. If we were cut off from all foreign oil tomorrow, we would be in serious trouble because we can't adjust that quickly. We can reach the goal of reducing foreign oil consumption by half within 10 years without too much difficulty. We are working on a multiplicity of technologies. Nobody knows which of these will be most successful. The mix will change as time goes on. If we have more use of solar energy, and that seems to be a field where growth is exceeding projections, it will reduce what we require from nuclear power, or coal, or some of the other technologies. As far as I am concerned, yes, we can easily correct this problem. The question is—what time frame?

*Robert L. Smith* — In reference to the comments on the environment and political solutions, the environment is too delicate to entrust either to the Democrats or Republicans, or anybody else as far as that goes. I don't even want to trust some environmentalists, really. So, that is a terrible question to answer. Yes, I'm optimistic, not very pessimistic. What I was really trying to point out was, you have to plan ahead and still have some room to maneuver; but we cannot continue on our present course, which is a crash course against future human welfare; and I'm as interested in human welfare as much as anybody else is. You would be crazy if you weren't, since we all belong to the same human species. My statement of no growth is subject to misunderstanding—you don't have time in 20 minutes to explain yourself fully. But I'm glad some people are thinking about it. We cannot continue on our present high-level energy consumption and growth based on that type of an energy consumption. We are running up against entropy. We do live in a finite world. I think we can work together—things are looking a little bit better already. I think the explanation is better environmental education for everybody. There was an article in *Science* a few weeks ago reporting an environmental survey and only 30% of the country knew what acid rain was. Now, if only 30% of the country knows what acid rain is, we're in trouble—after all, the information has been broadcast on TV, and reported on radio, newspapers, etc., etc., It is extremely important that everybody has some exposure to ecology and environmental education. I think all lawyers, engineers—in particular, all political scientists, all have a good course in ecology before they get their degrees. At the same time, all those who are environmentally oriented in college need a good course in economics, political science, sociology, technology and a few other things. Then we can begin to see other viewpoints, and we might work together, which we have to do. There have to be compromises made all along the line, and the sooner we come to a common frame of mind, the sooner we're going to solve the problems.

*Justice Flaherty* — Thank you very much and let me then take the opportunity to thank the members of the panel who have devoted their time, their efforts and a whole day. Certainly we'd like to thank those who came to participate in the symposium.

*Chapter Eight*

# Assessing The Impacts of Energy Production on Aquatic Ecosystems — One Company's Approach

**W.F. Skinner**[1]
**&**
**R.B. Domermuth**[2]
[1]Project Scientist and
[2]Environmental Scientist
PENNSYLVANIA POWER
& LIGHT COMPANY
Two North Ninth Street
Allentown, Pa. 18101

W.F. "Ric" Skinner is a Project Scientist in the Environmental Management Division of Pennsylvania Power & Light Company, Allentown, Pennsylvania. Ric has been with PP&L four-and-a-half years and was previously employed as an aquatic biologist with Consumers Power Company (Jackson, Michigan). Ric holds a B.A. (biology) from Bloomsburg State College and an M.S. (biology) from West Virginia University and has developed a broad background in environmental impacts of electric energy production.

Robert B. Domermuth is an Environmental Scientist in the Environmental Management Division of Pennsylvania Power & Light Company, Allentown, Pennsylvania. Prior to joining the company, Bob worked as a Fisheries Biologist for the State of Kansas and Research Biologist employed by Ichthyological Associates and Radiation Management Corporation. Bob holds a B.A. (biology) from SUNY-Pottsdam and an M.S. (Fisheries Biology) from the University of Massachusetts. Bob has developed an extensive background in both freshwater and estuarine aquatic studies.

*Abstract*

Electric utilities have in the past generally reacted passively to regulatory mandates regarding the assessment of environmental impacts of energy production on aquatic ecosystems. Several years ago, Pennsylvania Power & Light Co. began developing a new approach to respond to regulatory requirements, public concerns, and the company's environmental obligations. Two case histories, dealing with arsenic and thermal discharges, are presented in which PP&L initially played "follow the leader" with the environmental agencies and then moved ahead to assure that a complete and objective assessment was obtained in each situation. PP&L is expanding its voluntary efforts in aquatic impact assessment by establishing a bioassay laboratory and an environmental monitoring and surveillance program at its fossil-fired generating plants. This approach is strengthening the company's internal environmental awareness, developing cooperation and respect between the company and the agencies, and demonstrating that PP&L's role in environmental management is being taken seriously.

For those of you who are unfamiliar with Pennsylvania Power & Light Co., it is an electric utility which serves 29 eastern and central Pennsylvania counties. Annually, our 5 fossil-fueled and 2 hydroelectric stations produce approximately 33 billion-kilowatt hours of electricity for use by over 1 million residential, commercial and industrial customers. PP&L, like most industries, supplies an essential product (electricity), but sometimes not without some cost to the environment. We at PP&L are concerned about these environmental impacts for three basic reasons: one is because Federal and State laws *require* us to be concerned; the second is that you—the public—*want* us to be concerned; and, the third is that PP&L, as a user of environmental resources, *needs* to be concerned.

In the generation of electricity, environmental impacts can involve air, solid waste and water—and all must be addressed in day-to-day operations. It is the last category—water, or more specifically aquatic environmental impacts—that we would like to address.

The awareness of, and response to aquatic environmental impacts by electric utilities has, in general, paralleled changes in water pollution control efforts in the United States. Through the first half of the 20th century, these efforts were largely directed toward abatement of raw domestic sewage discharges into surface waters as a means of controlling water-borne infectious human diseases. Industrial wastes were not considered a major problem because during this period, prior to the rapid growth of the synthetic organic chemicals industry, industrial wastewaters were substantially more simple in chemical makeup. As America's industrial complex grew and changed in character, environmental awareness concerning pollutants and their impacts increased. Federal and state regulatory agencies shifted their emphasis from controlling problems associated with domestic wastes — such as oxygen depression, pH, and suspended solids — to identifying and controlling the more than 350 pollutants presently characterized as toxic or hazardous, including organic compounds and most heavy metals. Consequently, the aquatic environmental compliance requirements faced by industry have changed dramatically.

The 1970's marked the period of greatest change. With the passage of the 1972 Amendments to the Federal Water Pollution Control Act (FWPCA) and the later refinements known as the Clean Water Act (CWA) of 1977, a comprehensive Federal/state scheme was established for controlling the introduction of pollutants into the nation's waters. [For our purposes a pollutant can be defined simply as any substance (e.g., heavy metals), compound (e.g., organic chemicals) or effluent (e.g., thermal effluent) which is discharged in sufficient concentration or quantity so as to adversely affect the aquatic life in receiving waters (rivers and streams)].

The FWPCA Amendments provide for the creation of the National Pollutant Discharge Elimination System (NPDES) to control, by means of permits, the discharge of pollutant-containing wastewater. Under NPDES, effluent limitations are developed, placed in the "discharge permits" and enforced through monitoring requirements and the threat of stiff civil and criminal penalties to the discharger for non-compliance. Some effluent limitations are general in application and do not reflect site-specific conditions which may exist. In these instances, the limitations, if applied as mandated, could be more stringent then necessary to protect against water quality degradation. Therefore provisions are included in the law whereby a variance from a specific permit condition can be obtained for certain pollutants (e.g., heat). To obtain such a variance the discharger must demonstrate to the permit issuing agency that the permit condition is more stringent than necessary to protect the balanced resident community of aquatic life. Such a demonstration would normally consist of a scientific study developed at the direction of the regulatory agencies and paid for and conducted by the discharger.

The FWPCA was significantly refined by the Clean Water Act of 1977, which contains extensive references to the protection of aquatic life, stating that such protection can best be determined through a biological monitoring program. The law specifically addresses the identification of toxic substances and the prohibition of toxic discharges which may pose an unacceptable risk to humans in terms of health or the environment. It should be clear that this legislation does indeed *require* PP&L to be concerned about aquatic environmental impacts.

The second reason why we are concerned about these impacts is that you—the public—desire the maintenance of good water quality. Almost all the reasons for controlling or reducing water pollution are biological: fishing, boating, swimming, drinking water supply, agriculture and livestock watering. Each of these uses is directly important to everyone of you at one time or another, and you justifiably become concerned when any of these uses is potentially or actually compromised, especially by anthropogenic (man-induced) activities. Consequently, when these activities are PP&L activities, we may become concerned as well.

The third reason for PP&L to be concerned about aquatic environmental impacts stems from our obligation as a user of rivers and streams. The final step in the treatment of our industrial wastewater occurs in the State's surface waters. If this were not true there would be no need to discharge any wastewater because it could all be recycled and reused within the plant. However, this is not always technologically possible, economically feasible, or even in some instances ecologically justified. In fact, electric utilities in most situations (when operating normally) should not be considered major polluters of our waterways when compared with some other industries. However, the impact that we do have should not be under-emphasized either. Therefore, we as a Company, need to be actively aware of our environmental impacts.

Sandwiched among legal requirements, public concern, and industrial obligations, utilities have generally reacted passively to regulatory mandates. Environmental agencies have seized and maintained the initiative by identifying and prioritizing environmental concerns, dictating procedures to address these concerns, and determining compliance, all with minimal challenge or minimal input from the utility. In essence, the agencies have defined both the problem and the solution, leaving the utilities to play "environmental catch-up."

Several years ago PP&L recognized that a passive response was no longer adequate. We believed that a new approach was necessary to anticipate and address regulatory requirements, public concerns and the company's environmental obligations. We decided the time had come for PP&L to identify situations before they became environmental problems, to objectively assess the level of impact if it was occuring, and to cooperatively seek solutions with regulatory agencies rather than simply pursue solutions dictated by the agencies. This new environmental awareness did not spring up overnight. To the contrary, it evolved slowly during our experience in dealing with the environmental agencies.

Let's consider two examples in which the environmental agencies defined the problem *and* the solution and how PP&L, after initially "following the leader", recognized the need to go beyond the regulatory requirements to assure that a complete and objective assessment was obtained. Because the primary aquatic impacts of electric generation involve thermal and industrial wastewater discharges, our examples shall deal with these topics.

The first example deals with a discharge permit violation at our Montour Steam Electric Station. This plant is located in Montour County, Pennsylvania, on the upper drainage of Chillisquaque Creek, a small tributary of the West Branch of the Susquehanna River. Montour consists of two coal-fired units with a combined output of 1500 Mw. The plant utilizes two natural draft cooling towers, with makeup water being pumped from the West Branch of the Susquehanna River through a 12-mile pipeline. Plant effluent, which is a combination of cooling tower blowdown, flyash sluice water, coal pile runoff and miscellaneous plant waste waters, is discharged from the Industrial Waste Treatment Basin into Chillisquaque Creek. Plant effluent provides a substantial contribution to the creek—about 20:1 at the point of discharge.

In January 1976, the Pennsylvania Department of Environmental Resources (DER) cited PP&L for violation of the Clean Streams Law. Arsenic, at a concentration exceeding 0.7 ppm, had been measured by the DER in the Montour effluent. This was approximately 15 times the EPA criterion for treated drinking water and was alleged to be harmful to the aquatic life of the stream. The DER required PP&L to take action to abate the arsenic discharge. After our own testing confirmed the agency's finding, we hired several chemical engineering and analytical consultants to conduct wastewater characterization and treatment studies. These efforts, which continued through 1976, identified fly ash leachate as the source of the arsenic, however they were unable to explain just how it was being leached from the fly ash.

Effluent monitoring in 1977 indicated that arsenic levels had decreased substantially. Although the company has no official position on why levels decreased, it is thought to be due, in part, to the coincidental startup of a coal washing facility at the coal mine, which apparently resulted in the removal of much of the arsenic before the coal was shipped to Montour and burned. Nevertheless, in June 1977, about 18 months after the initial citation, the DER wrote the company:

> "As you recall, PP&L was initially notified of its discharge of arsenic in our letter of January, 1976. Since that time the arsenic concentration in the discharge has decreased from about 0.7 ppm to the present discharge level of 0.15 ppm. We understand this level is approximately that of discharges from other coal-fired generating stations. However, since it is discharged to a small stream, aquatic life is adversely affected and the public water supply uses of the stream are not protected."

We did not disagree that there was apparently some impact occurring immediately below the plant. However, we maintained that the impact was not as severe as DER alleged, nor was it specifically due to arsenic. We believed that DER's allegation was based on limited biological studies, most without good statistical controls and none of which were specifically designed to assess arsenic toxicity. Furthermore, there are no public water supplies on the stream, as alluded to in DER's letter, nor are any planned through the year 2000! On the other side of the coin, PP&L also had a limited biological data base and we could not objectively or confidently challenge the biological aspects of DER's allegations. We therefore entered negotiations for an engineering solution and an agreement on penalties. The DER-proposed remedy was to build a 12-mile discharge pipeline back to the Susquehanna River and to impose several millions of dollars in penalties.

We met with the DER and, after several discussions, mutually agreed that the

12-mile pipeline was neither economically feasible nor ecologically sound. However, we did commit to building a dry fly ash handling system, not so much to abate the arsenic discharge, which now averages less than 0.05 ppm (approximately the level allowable in treated drinking water), but because we believed this would be the best available technology for treating and reducing other pollutants, in addition to arsenic, in the industrial wastewater discharge at Montour.

The second thing we did was to undertake a number of voluntary programs to broaden our knowledge of arsenic impacts on aquatic life and more specifically, the effect of the Montour discharge on the biological and chemical aspects of Chillisquaque Creek. Shortly after we were first notified by DER of the arsenic problem, we hired aquatic biological authorities Dr. Max Katz, Dr. Chuck Wurtz and others to independently review the scientific literature providing us and the DER with a realistic perspective regarding arsenic impact on aquatic ecosystems.

We also began a two-year ichthyofaunal survey, under the direction of Dr. Bob Denoncourt, to document the distribution and relative abundance of fishes in relation to the Montour effluent. Concurrently, during the first year of the fish survey, we carried out an extensive water quality monitoring program with Gilbert Associates, analyzing for over 100 parameters in water and sediment samples. Finally, in order to address the DER concern about chronic toxicity of the effluent, we designed and built a bioassay laboratory for on-site studies with representative fishes. DER was provided study plans for its comments and suggestions, and it was kept up to date on these voluntary research efforts. Our efforts were instrumental in obtaining a successful resolution of the arsenic problem.

The second example concerns the thermal discharge from Brunner Island Steam Electric Station. This plant is located on the western bank of the Susquehanna River approximately 11 km below Harrisburg, Pa., and within eyesight of Three Mile Island. Brunner Island's three coal-fired units have a combined output of 1454 Mw. The plant utilizes once-through cooling, with two shoreline intakes and a shoreline discharge. The average temperature rise of water passing through the plant is 26 F (14.4 C) above the ambient river temperature.

In October, 1966, the State issued a thermal discharge permit which allowed the plant to discharge heated water so long as the temperature at the edge of the defined mixing zone did not exceed 93 F (33.9 C). Two years later, due to a change in State water quality standards, the maximum value was lowered to 87 F (30.6 C). The State requested PP&L to collect river temperature data to document the potential impact of the new limitation on plant operations. This monitoring effort continued for over 2 years. Our temperature data indicated the plant could not meet the 87 F limitation at all times. One way we could have achieved compliance with this limitation would have been to install cooling towers at a cost of $20-30 million. However, we believed this limitation was more stringent than necessary to protect the balanced resident community of aquatic life. So under provisions of Section 316 (a) of the CWA, PP&L applied to the Environmental Protection Agency (EPA) for a thermal variance. EPA required PP&L to demonstrate to EPA's satisfaction that granting of the variance (i.e., alternative thermal effluent limitations) was justified.

In 1976, PP&L agreed to conduct a 316(a) Demonstration, as such studies are called, and developed a study plan following EPA and DER recommendations. The Demonstration was to focus primarily on eight representative important species (RIS) of fishes selected by the agencies. At the agencies' direction, the Demonstration contained no 316(a)-specific field studies, but was basically a "desk-top" exercise because the agencies considered the existing data base was sufficient (to establish a successful demonstration). As a result, the 316(a) Demonstration report submitted in May 1977 was based primarily on literature values for thermal tolerance of RIS fishes, and computer modeling of the thermal plume utilizing plant operational and river data.

For the next 2½ years, during which time the State was granted permit authority, EPA indicated that the Demonstration was satisfactory, however, it was awaiting DER's review. In the interim, DER, in concert with the Pennsylvania Fish Commission (PFC) and the Susquehanna River Basic Commission (SRBC), requested three additional reports on specific topics. These were to address potential effects of Brunner Island operation on future migrations of American shad; probabilities of normal and worst case conditions in the River during critical periods for muskellunge spawning and smallmouth bass egg survival and development; and, an assessment of thermal effects based on a newly published method for deriving maximum weekly average temperatures.

The first two reports were "crystal ball" exercises in that American shad do not and probably will not occur in the near future near Brunner Island, because of the existence of three impassable hydroelectric dams downriver. Also, muskellunge are present only through stocking efforts and are not known to reproduce in the Susquehanna River. However, PP&L complied with the agencies' requests and generated the reports, again based upon the literature and available data.

In October, 1979, the company was surprised to learn that DER was going to recommend denial of the thermal variance request because PP&L had failed to adequately demonstrate to the DER that the thermal component of the cooling water discharge did not cause significant adverse impact to aquatic life in the River. When we were notified of the impending decision, we requested a meeting to discuss the basis of that determination. We believed our Demonstration was scientifically sound, and had hoped the reviewing agencies would also recognize the beneficial impacts, especially in relation to the recreational fishery which is enhanced by the thermal discharge. The meeting was worthwhile because it brought to light several misinterpreted aspects of the Demonstration. As a result, we were able to negotiate a resubmittal of the Demonstration with additional data, including specific field data and mathematical modeling studies to fill the gaps in the agencies' appraisal.

A revised 316(a) Demonstration, which included results of the 1980 field efforts, was recently submitted for agency evaluation. We believe this information will allow a more realistic appraisal of the effects of the Brunner Island cooling water discharge on aquatic life in the River.

Let me expand a little on one phase of the revised Demonstration that deals with positive impacts. We believe that the benefits provided anglers by the Brunner Island thermal discharge are an important consideration in any evaluation of impact on fishes. Several of the representative important fishes are game and pan species whose populations and availability to the angler appear to be affected by operations of Brunner Island. Our data show that while a life stage of some species may be excluded from the thermal plume during certain seasons, other life stages find refuge (preferred conditions) within the thermal plume and are therefore available to increased angling effort and success.

Angler surveys (creel census) were conducted in 1977-78 and again in 1980 to determine the recreational impact of the Brunner Island thermal discharge. Based on the data, we conservatively estimate that annually some 9,000 anglers enjoy 25,500 hours of fishing and catch 20,600 fish within the one mile of west shorezone immediately below the mouth of the thermal discharge channel. These estimates are over and above estimates for the east shorezone, which is not affected by the thermal plume. Most anglers surveyed were fishing purely for recreation and had no particular preference as to what species they might catch. We have made this information available to the environmental agencies reviewing our thermal variance request and hope they will realistically balance this beneficial impact against what our data show to be minimal adverse impacts.

PP&L has learned a great deal from our experience at Montour and Brunner Island. The two examples indicate that doing our homework and being prepared to respond to environmental concerns is necessary and invaluable. Consequently, PP&L is expanding its voluntary aquatic environmental programs in two areas. The first anticipates bioassay requirements in the next round of (NPDES) discharge permits by establishing a laboratory for acute bioassays of our power plant effluents utilizing a variety of aquatic organisms. While the electric power industry will probably not be blanketed with bioassay requirements, discharge permits for certain power plants, particularly where there has been a history of pollution incidents (e.g. Montour SES), will probably require bioassays.

The second area of voluntary effort will be to establish an environmental monitoring and surveillance program at all of our fossil-fueled power plants. In some ways, this will be a typical baseline program of monitoring for water quality, fishes and aquatic invertebrates on a seasonal basis. What makes this program unique is that, at our suggestion, it will be a cooperative effort with DER Regional biologists. We have already begun this program at the Montour plant.

The field team consists of DER and PP&L biologists working together to collect one set of data. Taxonomy and water quality analyses are shared, and taxonomic specimens cross-checked. One data report is prepared quarterly and an annual report with discussion and conclusions will be jointly authored.

We also plan to improve our pollution incident response procedures and incorporate them into this program for the timely investigation of significant aquatic pollution incidents, should they occur. The procedures will again include the PP&L/DER field team investigating the incident cooperatively and objectively. This may seem to be idealistic to some of you, however, from our efforts and discussions with DER to date we have excellent reason to believe that this program will work.

In conclusion, we believe the decision by PP&L to take a more active, constructive, and cooperative approach to assessing aquatic environmental impacts makes sense for several reasons:

- it is strengthening our internal environmental awareness and is providing a mechanism to address potential problems in the early stages.
- it is encouraging an in-house cost-effective technical expertise to be built-up
- it is providing a spirit of cooperation and respect between company and agency technical and administrative personnel
- it is demonstrating not only to the public and our own operating people, but to regulatory agencies as well, that Pennsylvania Power & Light Co.'s role in environmental management is being taken seriously.

*Chapter Nine*

# Surface Coal Mine Reclamation – A Need for Re-Evaluation

**Fred J. Brenner, Ph.D.**
Biology Department
GROVE CITY COLLEGE
Grove City, Pa. 16127

Dr. Fred J. Brenner received his B.S. degree in Biology from Thiel College and his M.S. and Ph.D. degrees from The Pennsylvania State University. He has been active in the Pennsylvania Academy of Science for many years and served as Editor of the Newsletter for seven years. In addition to his service to the Academy, he is a member of the Executive Council of the Ecological Society of America, District Director of Beta Beta Beta Biological Honorary and is currently serving as President of the Pennsylvania State Chapter of the Wildlife Society. Dr. Brenner is a member of the Mercer County Regional Planning Commission, serving as chairman of the Environmental Review Committee, Secretary-treasurer of the Mercer County Conservation District and has received the Award of Merit and Silver Beaver from the Boy Scouts of America for his work with youth. Dr. Brenner is a member of 16 professional societies and has published over 75 papers in professional journals, including numerous articles dealing with surface coal mine ecology, especially as it relates to fish and wildlife management. Dr. Brenner has been a member of the Biology Faculty at Grove City College since 1969.

*Abstract*

Current surface mine reclamation procedures are discussed in the light of ecological site conditions, plant succession and wildlife values. The practice of back grading to approximate the original contour may be detrimental to reclamation due to soil compaction and long slopes causing excessive erosion. The total biomass of vegetation was significantly correlated with the organic and moisture content of the soil. Volunteer species provide the major component of the biomass of herbaceous and forest communities. Volunteer species have a greater value as food and cover for wildlife than those used in initial reclamation. Aquatic areas should be encouraged where conditions permit because of their value as fish and wildlife habitats and in enhancing the genera ecology of the area. The use of Appalachian coal for energy and other industrial uses will increase in the next several decades. Surface or open-pit mining, where feasible, is less expensive and safer than deep or shaft mining. Criticism of the land upheaval and inadequate reclamation procedures of such mining operations has come from those segments of society concerned with environmental issues. In Pennsylvania alone, thousands of acres of land are disturbed and reclaimed annually with varying degrees of success. This paper endeavors to discuss current reclamation practices in light of ecological principles and proposes recommendations for restoration of these lands to their previous ecological diversity or even enhance them further as wildlife habitats.

*Introduction*

Odum (1) suggested that the disturbance of ecosystems can be managed in such a manner as to enhance productivity while maintaining a degree of environmental resiliency. Reclamation plans on disturbed lands should contain provisions for: (1) adequate sediment and erosion control; (2) habitat diversity that will provide food and cover for wildlife or agriculture production and (3) aesthetics while utilizing a reasonable cost benefit ratio (2,3). Unfortunately, current reclamation guidelines and criteria established by state and federal agencies are concerned primarily with establishing cover at the expense of diversity and wildlife habitat.

Brief History:

Surface mining began on a limited scale during World War I, although extensive mining did not begin in Pennsylvania until after World War II. In the early days, mining was generally restricted to coal seams 10 m (30 ft.) or less below the surface and little or no backfilling or revegetation was completed. The development of larger equipment in the 1950's and 60's made it possible to mine deeper coal seams over more extensive areas than was previously possible. Prior to 1963, reclamation consisted of limited backfilling and site preparation with pines (*Pinus* spp.), spruces (*Picea* spp.) and black locust (*Robina pseudo-acacia*) being planted. Current regulations in Pennsylvania are those promulgated under the surface mining act of 1963. Surface mine operators are currently required to backfill to approximately the original contour of the land and to re-vegetate with mixtures of various grasses and legumes. The passage of the Federal Surface Mining Act in 1977 provides two important provisions: (1) prime farm land soils that are designated by the U.S. Soil Conservation Service must be restored to productive farm land and (2) fish and wildlife habitats must achieve prime consideration. The inclusion of these provisions into a reclamation plan will involve the development of procedures specific for these objectives and will be discussed later in this article.

Site Preparation:

Current regulations require the removal and stockpiling of top soils (45.7 cm of soil) until the completion of the mining operation. At the termination of mining, soil overburden must be replaced in the same order as removed, the area regraded to approximately the original contour and top soil replaced and revegetated. Extensive backfilling and grading to original contour may be a deterrent to successful reclamation because of soil compaction and excessive wind or water erosion of top soil due to a lack of bonding of the soil layers. Moreover, grading to original contour often results in long and extensive slopes which accelerates erosion. The rate of soil erosion is directly proportional to the length of slope, not the degree of slope (4). A series of short slopes or terraces along with water diversions along the contours would decrease erosion and accelerate the establishment of vegetation on the site. Other problems created during site preparation and top soil replacement included: (1) difficulty in obtaining an even distribution of top soil on the site; (2) lack of fertile top soil because of prior farming or land use practices, and (3) possible changes in the physical, chemical or biological properties of the top soil because of extensive storage time. Top soil, however, is not essential for quality reclamation providing that the species used in reclamation are suitable to site conditions.

Soil Characteristics:

Surface mine soils exhibit considerable variation due to a combination of previous land use, original soil composition and backfilling practices after mining. A survey of 81 surface mine sites indicated a pH range varying from 3.5 - 8.5 with the majority between 4.5 - 5.5 (Fig. 1). A total of 16 different soil parameters

TABLE 1

*Correlation Coefficients of Soil Parameters as Determined by a Matrix Correlation Analysis.*

| | Percent Organic Matter | Percent Soil Moisture | Soil Moisture at Saturation | ph |
|---|---|---|---|---|
| Percent Organic Matter | 1.000 | 0.750 | 0.570 | 0.791 |
| Percent Moisture | 0.752 | 1.000 | 0.804 | 0.605 |
| Water Holding Capacity | 0.871 | 0.804 | 1.000 | 0.894 |
| pH | 0.791 | 0.604 | 0.894 | 1.000 |

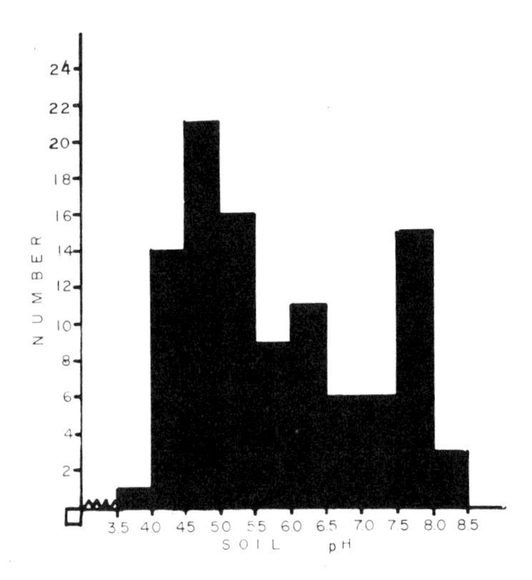


FIGURE 1. Distribution of soil pH on 81 surface mines in Mercer County, Pennsylvania.

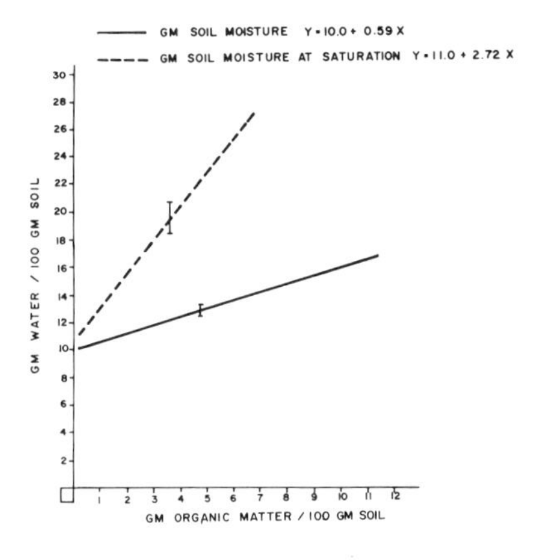


FIGURE 2. Relationship between the amount of organic matter and soil moisture concentrations in surface mine soils. (soil moisture r 0.850; $P < 0.001$; at saturation r 0.888; $P < 0.001$).

were measured and the concentrations of micro- and macro-nutrients did not vary significantly among the different sites. Surface mines generally possessed a total available nitrogen and potassium concentration of 11.4 kg/hectare (10 lb/acre) or less per hectare and a phosphorus content of 113.6 kg/hectare (100 lb/acre) or less. The success of reclamation is, however, significantly correlated with two soil parameters: organic content and moisture. Soil moisture and the amount of moisture that soil can retain at saturation are significantly correlated with organic content (Table 1, Fig. 2). In both Pennsylvania (5) and West Virginia (6), the biomass of vegetation occurring on surface mines was significantly correlated with the organic content. Studies in Pennsylvania (5) demonstrated that the biomass of vegetation was significantly correlated with organic content (Fig. 3) and moisture (Fig. 4), with pH being of minor importance (Table 2). The relationship between soil characteristics and vegetation indicate that improved reclamation methods should be utilized to increase the organic content of the soil. The use of inorganic fertilizer and lime will aid in the early establishment of grasses and legumes but they are of little value in the survival and growth of

TABLE 2

*Correlation Coefficients between soil parameters and plant characteristics on surface mines in Mercer County, Pennsylvania (5).*

| Plant Characteristics | Percent Organic Matter | Percent Soil Moisture | Soil Moisture at Saturation | pH |
|---|---|---|---|---|
| Biomass/gm/m$^2$ | 0.654 | 0.745 | 0.322 | 0.319 |
| Height — cm | 0.375 | 0.136 | 0.245 | 0.102 |
| Density | 0.134 | 0.214 | 0.604 | 0.337 |

vegetation over an extended period of time. Organic materials, on the other hand, are generally longer lasting, aid in water retention and provide a media for the bacterial decomposition of dead material, thereby increasing the organic content of the soil. Organic matter also aids in the establishment and growth of trees and shrubs. For example, quaking aspen (*Populus tremuloides*) grew 0.955 m/year after a single application of sewage sludge compared with 0.12/year for an adjacent control area. Likewise, bristley locut (*Robina hispida*) had an average growth of 0.36 m/year while those on the control site grew 0.088/m (5). Continual application of sewage sludge may result in the accumulation of heavy

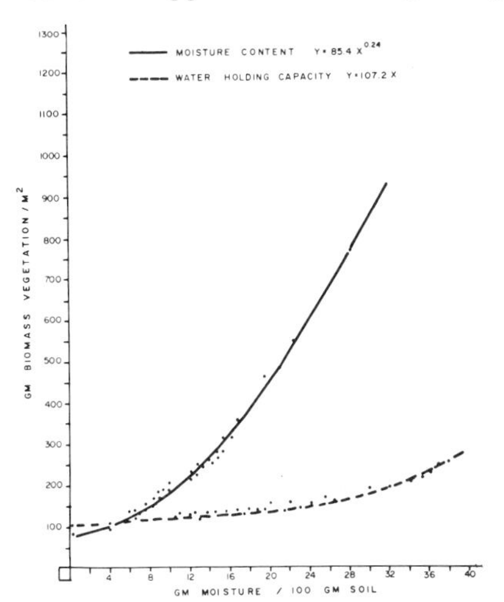


FIGURE 3. Relationship between soil moisture concentrations and the biomass of vegetation present on surface mines (soil moisture r .872; P < 0.001; at saturation r .810; P < 0.01)

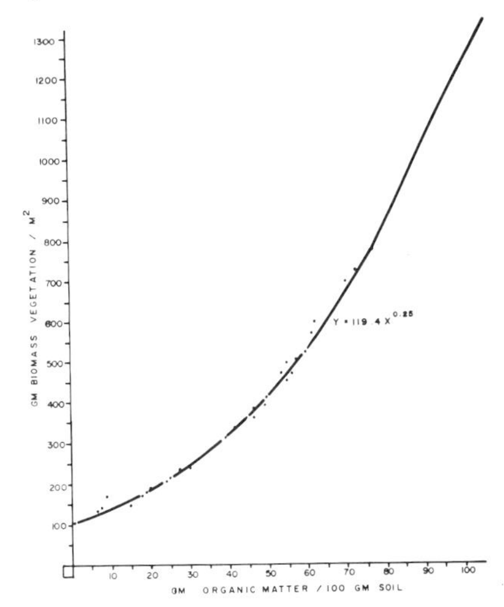


FIGURE 4. Relationship between the amount of organic matter in the soil and the biomass of vegetation present on surface mines (r .863; P < 0.001).

metals (7) but a single application in most cases is sufficient to provide an adequate media for plant growth.

Evaluation of Vegetation Associations:

An important component of herbaceous communities on surface mines is the number and contribution of volunteer species that invade the area. The success of reclamation efforts may, therefore, depend on the creation of a suitable habitat for the invasion of volunteer species. Thirty-two different species of plants were identified on 10 surface mines representative of western Pennsylvania with only 5 or 15.6 percent of these species being utilized in the initial reclamation. The importance values of these plants were evaluated on the basis of their contribution to total biomass, chlorophyll *a* concentrations and their persistance over a period of years. Each value was expressed as a percentage of the total; hence, for a pure stand of single species, the importance value would be 300.

The total importance value for volunteer species on these mines varied from 31 to 93 percent. However, if only biomass was used as an indication of importance, volunteer species varied from 0.5 to 82 percent of the total biomass. The ratio of volunteer/reclaimed species was a determining factor in both the portion of the total biomass and the overall importance of volunteer species in the community (Fig. 5). A dense cover composed of species utilized in reclamation (low vol./reclam. ratio) would provide an environment unsuited to the invasion of volunteer species, thereby reducing diversity on the area.

The selection of species for the initial reclamation may also be a factor in determining the number and biomass of volunteers invading the area. For example, a grass, clover or birdsfoot trefoil (*Lotus corniculatus*) mixture generally will become established in sufficient amounts to prevent massive erosion but allowing for invasion of volunteer species. On the other hand, crown vetch (*Coronilla varia*) is difficult to establish but provides a dense cover that inhibits volunteer invasions. The use of sorgum (*Sorgum vulgare*) as an initial cover crop appears, on the basis of preliminary trials, to provide the best of two worlds. On the one hand, it provides a dense cover preventing erosion and encourages the invasion of volunteer species. Stands of sorgum were used extensively by morning doves (*Zeneidura macroura*) and other species which may have aided in the establishment of volunteers by the passage of seeds through their digestive tract or other accidental transfers. The coverage provided by volunteer species the following year was 6 percent greater than it was on adjacent control planted the previous

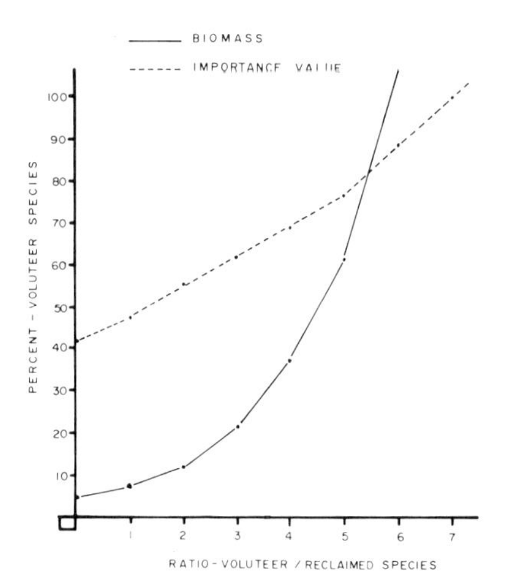


FIGURE 5. Relationship between the ratio of volunteer species to those used in initial reclamation and the percent of volunteer species in the total biomass.

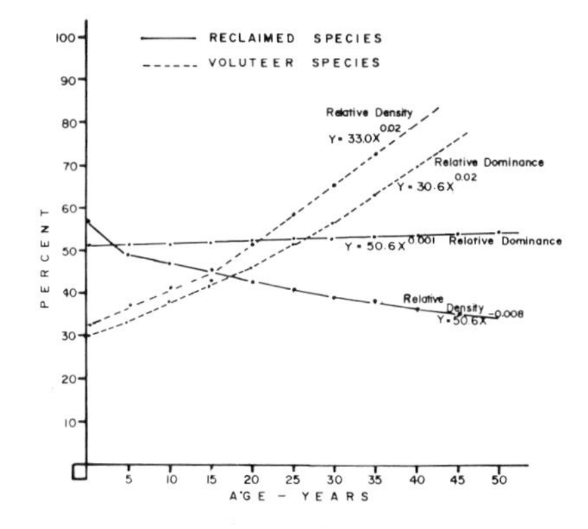


FIGURE 6. Changes in relative dominance and relative density of volunteer species and those used in initial reclamation through time in forest communities on surface mines in Mercer County, Pa.

year with grasses and legumes. The development of a forest community on surface mines is similar in many respects to that of a grassland community. A total of 34 different tree and shrub species were identified on surface mines and, of these, 22 were volunteer species. Botkin (8, 9, 10) indicated that the importance of a species in a forest community may be determined by a relationship between the diameter (DBH), height and leaf area. A combination of basal area (relative dominance), relative density, canopy height and relative growth factor (Botkin, 8, 9, 10) was used to determine the overall importance of individual species to the community. In a pure stand, the importance value would therefore be 400. In general, volunteer species increase in importance to the with age but the time sequence will vary, depending on which parameter is being evaluated. Relative density of volunteer and reclaimed species are approximately equal 15 years after mining, whereas the relative dominance of the two groups are not equal until approximately 30 years after mining. The overall importance of the two groups are equal after 25 years and then reclaimed species decline in importance with a corresponding increase in volunteer species (Fig. 6). Aspen (*Populus tremuloides, P. granidentata*), red maple (*Acer rubrum*), black cherry (*Prunus serotina*) and several species of oaks (*Quercus* spp.) and hickories (*Carya* spp.) provide the greatest contributions to forest communities on surface mines. In general, a dense biomass of herbaceous vegetation is detrimental to the establishment, growth and development of forest communities on mined lands (Fig. 7) (5). However, preliminary studies indicate that it is possible to direct seed apples (*Pyrus malus*), hawthorne (*Crataegus* spp.), oaks, hickories and walnuts simultaneously with grasses on surface mines, thereby providing an initial cover to prevent erosion while providing long term benefits of establishing a forest community.

Wildlife Benefits:

Surface mines in Pennsylvania (2, 3, 11, 12, 13, 14), Ohio (15), Illinois (16) and West Virginia (17, 18, 19, 20) have been shown to have considerable potential as wildlife habitats. These benefits, however, are primarily a result of natural succession of native wildlife food and cover species rather than reclamation procedures designed to benefit wildlife. In general, current regulations and reclamation practices are detrimental to wildlife because they are creating a mono-culture lacking diversity. This occurs, regardless of the fact that state and federal regulations indicate that surface mines may be reclaimed for wildlife.

Studies conducted on surface mines in Pennsylvania found 54 species of birds and a number of mammalian species using mines on either a permanent or semi-permanent basis (migrant birds). Both avian and mammalian communities showed a greater correlation with volunteer species than with those used during reclamation. Moreover, studies on the nutrient content and use of these species by wildlife indicated that feeding activity was greater on volunteer species with higher nutrient content than it was on exotic species used in reclamation (3). This

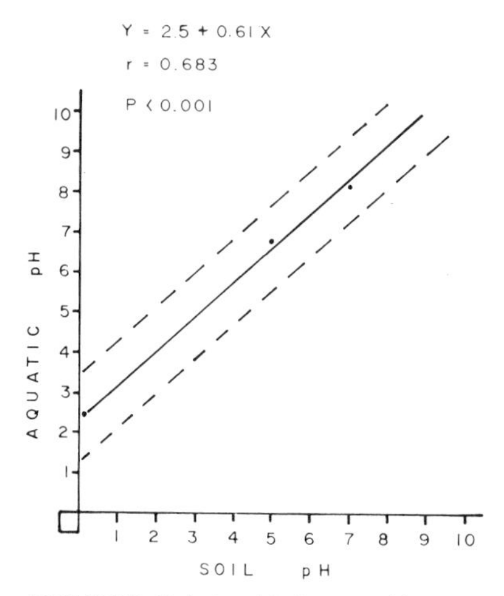


FIGURE 7. Relationship between biomass of herbaceous vegetation and basal areas of forest communities.

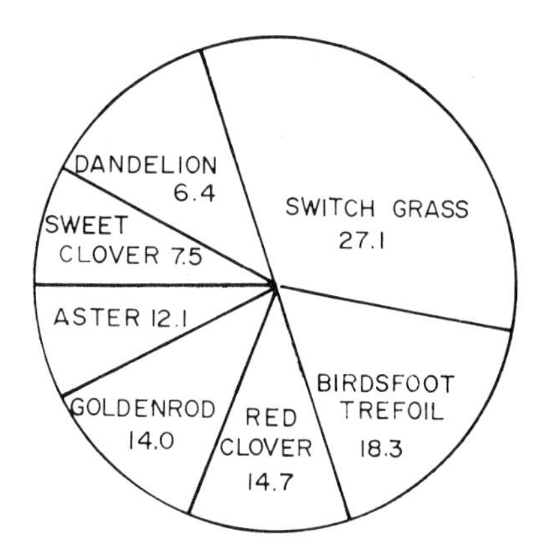


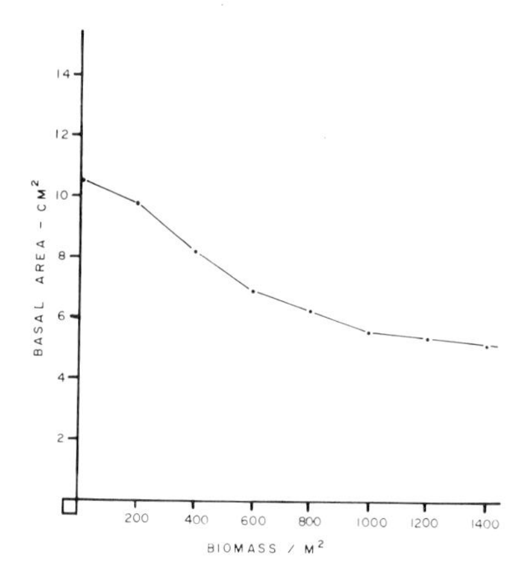


FIGURE 9. Relationship between the soil pH and the pH of surface mine pools based on 132 pools located on 81 different surface mines in Mercer County, Pa.

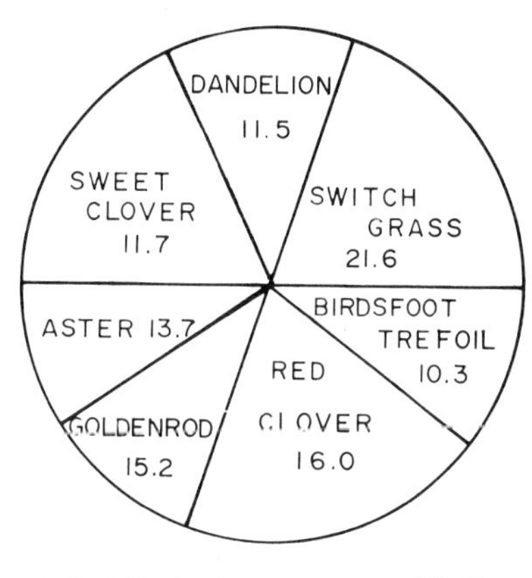


FIGURE 8. Comparison of relative importance values of seven species of plants found on a hypothetical surface mine area including their value as cover as well as their value for wildlife.

is further illustrated by a comparison of the increase in importance value of 6 species and the decline of others when the total digestible nutrients were included in the calculations (Fig. 8). The inclusion of species such as sorgum, red clover and those species indigenous to the area and the elimination of fescue and birdsfoot trefoil would benefit wildlife. Surface mines may be further enhanced by including tree and shrub species such as bristley locust, autumn olive (*Elaeagnus umbellata*), dogwood (*Cornus* spp.) and honey suckle (*Lonicera* spp.) that have food value for wildlife as well as value in erosion control as stated previously.

Primary results indicate that most producing trees and shrubs can be directly seeded along with an initial seeding of grasses and legumes, thereby insuring long term wildlife benefits. These species may be planted along the contours as hedgerows interspersed with pines and spruces for cover which would also provide wind brakes, erosion barriers and food and cover for wildlife at relatively low cost.

Aquatic Aspects:

Aquatic habitats on surface mines provide ecological benefits that should be considered a vital segment of reclamation planning. These areas function as sediment traps, allow for recharge and release of ground water, increase fish and wildlife habitats and provide a basis for future water supplies and recreation. In general, the pH of the pool is directly proportional to the pH of the surrounding spoil (Fig. 9). A survey of 132 pools located on 81 different mines in Mercer County, Pennsylvania indicated that only 11 would not support aquatic life (12). These areas supported 99 different taxa of phytoplankton and zooplankton (13). In addition to the different species of panfish (*Lepomis* spp.), yellow perch (*Perca flavescens*) and largemouth bass (*Micropterus salmoides*), these areas also supported walleye pike (*Stazostedion vitreum*), northern pike (*Esox lucius*) and tiger muskelunge (*Esox masquinongy* X *E. lucius*) as well as other species (13). These areas are especially beneficial to waterfowl (21, 22) and the addition of artificial nesting structures will increase breeding populations (21). These areas may be further enhanced by planting wildlife food species along the shorelines (11). Space does not permit a detailed discussion of these unique habitats but it is sufficient to state that they should be included into a reclamation plan where site conditions and water quality permit their inclusions.

Recommendations:

1. Substitute reclamation practices based on ecological and site conditions and not on agricultural practices, except where farming is the primary objective.
2. Do not back blade the site, avoid excessive compaction and rough up the site with a harrow prior to planting.
3. Eliminate long slopes and install water diversions along contours.
4. Use species in the initial reclamation that provide food and cover for wildlife, stabilize the area, encourage natural selection and eliminate exotic species.
5. Encourage the planting of trees and shrubs that provide food and cover for wildlife along contours either by direct seeding or conventional planting.
6. Encourage the establishment of aquatic areas wherever conditions permit.
7. Develop regulations that are results oriented rather than technique oriented. Quality reclamation should be the primary objective rather than what techniques or species are used during reclamation.

## ACKNOWLEDGEMENTS

Studies reported in this paper were supported by the National Geographical Society and the Office of Water Resources Research, U.S. Department of Interior. Thanks are expressed to Mr. Richard Crowley for his help in several aspects of the field work and the critically review of various stages of the manuscript.

## REFERENCES

1. Odum, E.P. 1969. The study of ecosystem development. Science 16: 262-270.
2. Brenner, F.J., R.H. Crowley, R.H. Musaus and J.H. Goth III. Evaluation and recommendations of strip mine reclamation procedures for maximum sediment-erosion control and wildlife potential. 3rd Symp. on Surface Mining and Reclamation. 11:3-23.
3. Brenner, F.J. 1978. Food and cover evaluation of strip mine plants as related to wildlife management. pp. 294-305. *In* D.E. Samuel, J.R. Stauffer, C.H. Hocutt and W.F. Mason, Jr. (eds). Proc. Surface Mining and Fish/Wildlife Needs in the Eastern United States. FWS/OBS - 78/81 386 pp.
4. Spangler, M.G. and R.L. Handy. 1973. Soil Engineering. Intext Educational Publishers. New York. 748 pp.
5. Brenner, F.J. 1978. Soil and plant characteristics as determining factors in site selection for surface coal mine reclamation. 1:37-44.
6. Staples, J. 1977. Vegetational succession, soil characteristics, and primary production and energetics on surface mines. Unpub. Master's Thesis, West Virginia University, Morgantown. 134 pp.
7. Goodman, G.T., C.E.R. Pitcairo and R.P. Gemmell. 1973. Ecology affecting growth on sites contaminated with heavy metals. pp. 149-172. *In* Ecology and Reclamation of Devastated Land. Gordon and Breach. New York. 504 pp.
8. Botkin, D.R. 1972. Some ecological consequences of a computer model for forest growth. J. Ecol. 60: 840-872.
9. Botkin, D.R., J.F. Janak and J.R. Wallis. 1972. Rationale, limitations, and assumptions of a northeastern forest simulator. IBM J. of Research and Development. 16: 101-116.
10. Botkin, D.R. and R.S. Miller. 1974. Complex ecosystems: models and predications. American Scientist. 62: 448-451.
11. Brenner, F.J. 1973. Evaluation of abandoned strip mines as fish and wildlife habitats. Trans. N.E. Wildlife Conf. 30:205-229.
12. Brenner, F.J. 1974. Ecology and productivity of strip mine areas in Mercer County, Pennsylvania. Research Technical Completion Report A-029-PA.

Institute for Research on Land and Water Resources. The Pennsylvania State University. 70 pp.

13. Brenner, F.J. 1978. Evaluation of factors promoting the preservation of aquatic ecosystems in reclaimed strip mine areas. Research Technical Completion Report. A-044-1A. Institute for Research on Land and Water Resources. The Pennsylania State University. 62 pp.
14. Brenner, F.J., M.J. Musaus and W. Granger. 1977. Selectivity of browse species by white-tailed deer on strip-mine lands in Mercer County, Pennsylvania. Proc. Pa. Acad. Sci. 51: 105-108.
15. DeCapita, M.R. and T.A. Bookhout. 1975. Small mammal populations, vegetational cover and hunting use of an Ohio strip-mine area. Ohio J. Sci. 75: 305-313.
16. Klimstra, W.B., P.A. Vohns, Jr. and J.D. Cherry. 1963. Strip mine lands for recreation. Illinois Wildlife. 18: 1-6.
17. Wray, T., III, P.B. Wackenhut and R.C. Whitmore. 1978. The reproductive biology of passerine birds breeding on reclaimed surface mines in northern West Virginia. pp. 333-344. *In* D.E. Samuel, J.R. Stauffer, C.H. Hocutt and W.F. Mason, Jr. (eds). Proc. Surface Mining and Fish/Wildlife Needs in the Eastern United States. FWS/OBS 78/81 388 pp.
18. Kimmel, R.O. and D.E. Samuel. 1978. Ruffed grouse use of a twenty year old surface mine. pp. 345-351. *In* D.E. Samuel, J.R. Stauffer, C.H. Hocutt and W.F. Mason, Jr. (eds). Proc. Surface Mining and Fish/Wildlife Needs in the Eastern United States. FWS/OBS 78/81 388 pp.
19. Brown, S.L. and W.E. Samuel. 1978. The effects of controlled burning on potential bobwhite quail brood habitat on surface mines. pp. 352-358. *In* D.E. Samuel, J.R. Stauffer, C.H. Hocutt, and W.F. Mason, Jr. (eds). Proc. Surface Mining and Fish/Wildlife Needs in the Eastern United States. FWS/OBS 78/81 388 pp.
20. Whitmore, R.C. 1978. Managing reclaimed surface mines in West Virginia to promote nongame birds. pp. 381-388. *In* D.E. Samuel, J.R. Stauffer, C.H. Hocutt and W.F. Mason, Jr. (eds). Proc. Surface Mining and Fish/Wildlife Needs in the Eastern United States. FWS/OBS 78/81 388 pp.
21. Brenner, F.J. and J.J. Mondok. 1979. Waterfowl nesting rafts designed for fluctuating water levels. J. Wildlife Manage. 43: 979-982.
22. Sandusky, J.E. 1978. The potential for management of waterfowl nesting habitat on reclaimed mined land. pp. 325-327. *In* D.E. Samuel, J.R. Stauffer, C.H. Hocutt and W.F. Mason, Jr. (eds). Proc. Surface Mining and Fish/Wildlife Needs in the Eastern United States. FWS/OBS 78/81 388 pp.

*Chapter Ten*

# Environmental Enigma of Energy Generation with Solid Fossil Fuels

**James P. Miller, Jr., Ph.D., PE**

Associate Professor of
Civil Engineering
UNIVERSITY OF PITTSBURGH
Pittsburgh, Pa. 15261

Dr. Miller's interests are in the economic and administrative aspects of Water Resources and Environmental Engineering. His current research interests are involved with optimum planning for cost effectiveness in sewage treatment facilities, and the impact of operator training programs on environmental control. Dr. Miller serves as Chairman of the Commonwealth of Pennsylvania Board for the Certification of Water and Sewage Treatment Plant Operators. He is President and Chairman of the Public Works Committee of his borough council. For these activities he was named one of the Community Leaders of American in 1970.

Dr. Miller was elected to the Water Pollution Control Association of Pennsylvania Board of Directors and is serving as editor of the Water Pollution Control Association of Pennsylvania magazine.

## ABSTRACT

Energy generation with petroleum and natural gas was on the increase until the petroleum and natural gas shortages began in the mid seventies. Coal was on the decline as a source of energy because of its greater problems with air pollution, water pollution and solid waste production. As the major available fossil fuel, the shift to coal as an energy source gives impetus to solve the related environmental issues.

Air, water and solid waste environmental issues are all interrelated. When impurities are removed from air, they then become either a water impurity or a disposal problem as a solid waste. This is also true for impurities removed from water. If incinerated, they become an air pollution consideration, and if disposed of as a solid waste, the contaminants can leach back into the water system again.

The issue of air quality and the problems with sulfur oxides, particulates, carbon monoxide, nitrogen oxides and hydrocarbons are considerable. These oxides and their relationship and control of acid rain are discussed.

The water and solid waste problems that arise from the various desulfurization processes are compared. Also the environmental impacts from the mining of coal are discussed.

Until alternate sources of energy can be developed, coal and nuclear energy are the primary sources of energy for power generation in the near future. The accompanying environmental enigmas have yet to be solved.

## INTRODUCTION

Thirty years ago the trend in the electric power industry was to convert to oil and natural gas for fuel. The trend was away from coal as a fuel because of its related environmental problems. Air pollution from emissions, water pollution and the disposal of solid wastes all made it more economical to shift to oil or natural gas. The oil embargo in 1973 has brought not only the United States, but the world to realize its over dependence on petroleum and natural gas for energy production.

In the United States, maximum use of coal occurred in 1910 and then gradually decreased as petroleum and natural gas took over as major energy sources[1]. Utilizing nuclear energy as an energy source started in the mid 1960's and is increasing. Toledo Edison Co. in their 1980 annual report states that their net generating capability is 25 percent nuclear, 69 percent coal, and 6 percent other.

The annual growth rate of energy consumption in the United States today is 2.8% down from 4.1% in the 1960-73 period. The world rate is 3.1%. Many increases in environmental problems can be associated with this demand for more energy.[3–6]

It has been possible to estimate the amount of energy available from most known and economically recoverable sources. For instance, coal could serve as an energy source for 169 years if all the energy consumed would come from that source alone. However, solar radiation has in one year four times the energy of all coal.[7] The main problem is one of harnessing the sun's rays since they are so dispersed over the earth's surface. Also, they are easily disrupted by cloud cover and configuration.

Nuclear fission as a source of energy used in today's conventional reactors has a depletable life of 25 years. This is almost the same as the sum of the depletable life of petroleum and natural gas, which is 22 years.[7] Nuclear fission also has its unsolved environmental problems with control of radioactive emissions and disposal of radioactive waste.

Until alternative sources of energy can be developed, coal remains the primary source of energy. As breeder technology, nuclear fusion, geothermal heat, oil shale processes, solar radiation, windpower, hydropower, etc. are developed, they may play an ever increasing role in power generation. One must also keep in mind the time lag between the development of economic and feasible alternatives and their use.

Of the total reserve of $7.4 \times 10^{12}$ metric tons of known coal reserves in the world, the United States and Russia have 77%. Western Europe, a high user of energy, has only 5%.

Coal as an energy source has its problems. It is a dirty fuel to burn. Particulate, nitrogen and sulfur dioxide emissions are three major concerns. The disposal of the ash or residue from the burning processes poses another dilemma.

One should also point out that coal, being a solid fuel, has more material handling difficulties than liquid or gaseous fuels. Imagine the running of an automobile by a steam engine into which coal has to be shoveled. The conversion by the railroads from steam to diesel engine further proves this point.

For the stationary power generation sources, such as an electric utility, coal on the present horizon is the leading source of fuel for power. When liquid or gaseous fuels are required, coal will be a major contributor. Coal gasification and liquid conversion can become an economic feasibility when a liquid or gaseous fuel is required.

Both the liquid and gaseous conversion processes and the direct burning of coal have their environmental enigmas. The air pollution, water pollution, and solid waste disposal problems are ever present and greatly add to the cost of using coal. These problems are not independent but are interdependent. For example, the sulfur emissions in the form of sulfur dioxide can increase the acidity of rainfall to effect the pollution of our lakes and streams. Likewise the heavy metals and other hazardous constitutents of the solid wastes can leach out into our water resources.

*Air Quality*

The Clean Air Act of 1970 set a goal of attaining health-protecting primary standards by July 1975. This goal was not reached universally. In 1977, Congress set new deadlines in the Clean Air Amendments. The standards were to be achieved as expeditiously as possible but not later than 1982. 1987 is the deadline for carbon dioxide and photo-chemical oxidants (ozone) which are two pollutants closely associated with the transportation industry.

The Environmental Protection Agency has to date set standards for the following pollutants: sulfur oxides, particulates, carbon monoxide, nitrogen oxides, hydrocarbons and lead. Stationary fuel combustion processes are responsible for 21.9 million metric tons of sulfur dioxide or 81.4% of the total emission of pollutants in the U.S. Of the nitrogen oxides, 11.8 tons or 51.3% of the total emission is emitted from stationary fuel combustion processes.[8] 4.6 out of 13.4 tons or 34.3% of particulate emission of less than 20 microns but greater than 3 microns in size, is from stationary sources while 1.3 out of 5.0 or 16.0% of the smaller particles come from the same source.[9]

Of the total carbon monoxide pollution, 2.4% come from stationary sources. Five percent of the total hydrocarbon are emissions from this same type of source. Carbon monoxide and hydrocarbons from stationary sources do not cause a high degree of pollutant concern.

Sulfur oxides are emitted from the burning of sulfur containing fuels. Sulfur in the form of sulfur dioxide comprises about 95% of these emissions. Sulfur dioxides are associated with many respiratory diseases such as coughs and colds, asthma, bronchitis and emphysema. While these physiological effects are serious, many persons concerned with the environment feel the sulfur compounds such as sulfuric acids and sulfate salts are the most harmful.

Each of the various types of coals contain different amounts of sulfur. Bituminous coal, which makes up 46% of the coal reserves has the greatest sulfur content. Over 70% of this bituminous coal has a sulfur content of over one percent.[10]

United States air quality standards set the upper limit for the amount of sulfur dioxide that can be emitted per kilo calorie of coal burned in a power plant at 1.2 mg. The quick conclusion is that the western coals with their low percentage of sulfur have a marked advantage. However, their low percentage of sulfur is offset by their low heating value.

*Solid Wastes*

To reduce or eliminate sulfur emissions, the sulfur has to be either removed or reduced before the coal is burned or from the flue gas after it is burned. Two processes can be used to accomplish the removal of sulfur. One is the fluidized bed combustion (FBC) process. The other is the flue gas desulfurization (FGD) system.

The fluidized bed combustion process (FBC) is one in which the coal is burned in a fluidized bed of limestone or dolomite sorbent.

The flue gas desulfurization (FGD) system commonly used for removal of sulfur from flue gas is a scrubber system that accomplishes the removal of sulfur dioxide by contacting the gas with an alkali-containing sorbent such as lime or limestone. Scrubber techniques include venturis, spray tower, impingement plates, packed towers or turbulent contractors. The lime or limestone reagent reacts with the flue gas and the resultant residual consists of fly ash, calcium hydroxide, calcium carbonate, calcium sulfate, calcium sulfide and water.

Environmental assessment for both the (FGD) and (FBC) systems show great amounts of solid waste. Fly ash is a product of both systems. The leachates from both wastes contain large amounts of calcium sulfates and total dissolved solids. The (FBC) system had the higher pH values. Trace elements are higher in the (FGD) waste leachates due to the greater amount of fly ash present. Very little polycyclic aromatic hydrocarbons (PAH) were found in both the (FBC) solid waste and fly ash. Little data is available for (PAH) in the (FGD) sludges. Biological testing showed that the spent sorbent from the (FBC) units had some mutagenic effect on soil organisms. The (FBC) leachate had toxic effects on both freshwater and soil organisms. The fly ash also showed mutagenicity and toxicity for all studies.[11]

Subtitle C of Public Law 94-580 enacted in 1976 is known as (RCRA) the Resource Conservation and Recovery Act. This law imposes strict controls over the management of hazardous wastes through its entire life cycle. The wastes from both the (FBC) and (FGD) processes are good candidates to be classified as hazardous wastes because of the exhibition of mutagenic and toxic properties. Furthermore, the ash concentrations from both processes can exceed the RCRA Guidelines.[12]

It is not conclusive which process (FBC or FGD) has the greatest solids generation because the range of the amount of solids from the (FBC) process is so great.[14]

The FGD process uses water, and the solids generated are difficult to remove economically; therefore, less sludge might be expected from the FBC process.

Many of the metals in the leachates from the FBC and FGD solid wastes are within the drinking water standards. As expected, both the calcium and sulfate are well above the standards.[15,18]

In the production of synthetic fuels from coal, sulfur is removed as a part of the process. Limestone or lime is again reacted with the sulfur oxides or hydrogen sulfide. The sulfur waste products formed are much the same as in the FGD and FBC processes.

The energy consumed in conversion of coal into gas or liquid fuel makes the synthetic fuel process less efficient than the direct burning process in overall energy output. However, because the sulfur is removed quite easily in the conversion process, it is quite likely that the conversion processes may be the most economical for the use of high sulfur coals.

*Nitrogen Oxides*

Nitrogen oxides are formed when any fuel is burned at a high enough temperature. Some of the 79% nitrogen in air will react with 21% oxygen in air at temperatures above 650 °C or 1200 °F. Nitrogen dioxide is the most plentiful of these oxides. $NO_2$ is now the oxide that is monitored after a reliable technique for its measurement was approved in 1976.

The principal source of nitrogen oxide emissions are electric utility and industrial boilers (56%) and auto and truck engines (40%).

Oxides of nitrogen can irritate lung tissue causing bronchitis and pneumonia as well as lowering resistance to such infections as influenza. These oxides of nitrogen also react with water vapor to form nitric acid, which is a constituent of acid rain. Also, the oxidants that these oxides of nitrogen help form in the presence of hydrocarbons and sunlight are the primary ingredients of photo chemical smog.

The burning of coal is relatively innocent as to the amount of unburned gaseous and vaporous fuels in the air known as hydrocarbons. Most of the 28 million metric tons emitted each year in United States comes from automobiles, industries that use solvents, and the users of paints and dry cleaning fluids.

Peroxyacetal nitrates (PAN) are formed by the chemical reaction in air of the nitrogen oxides and the unburnt hydrocarbons. Sunlight is needed for this reaction, hence, the title of photochemical smog. This smog, together with ozone, has a deleterious effect on persons with respiratory ailments and causes eye irritation as well as extensive damage to plant life.

Carbon monoxide is another air pollutant, but only 1.2 million tons out of a total of 87.2 million tons (1.28%) comes from stationary fuel combustion processes.

*Acid Rain*

When the oxides of nitrogen and the oxides of sulfur combine with water, nitric and sulfuric acids are formed.

The weighted annual average of pH of precipitation in the eastern United States in the periods 1955-56 and 1972-73 shows a significant lowering although one might expect an increase with increased use of solid fossil fuels.

Carbon dioxide plays a role in decreasing the pH of rainfall.[19] A good average value for carbon dioxide in the atmosphere is about 325 ppm. Therefore, rain water saturated with atmospheric carbon dioxide would have a carbonic acid concentration of $10^{-5}$ mols/1 and a pH of 5.7. The concentration of carbon dioxide in top soil is about 100 times that of the air with a resultant pH of 4.7. After passing through the humus layer of the soil, the rainfall water will react with the limestone, silicate clays, and other minerals with a resultant increase in pH. Many lakes have a pH between 7 and 8. The pH of the oceans is about 8, slightly on the alkaline side.

When the pH of precipitation starts to fall below 4.7 and even less than 4.0, one can look to sources such as sulfuric and nitric acids in the atmosphere. Of the ions present in rainfall measured at Hubbard Park, New Hampshire in 1974, 69% of the anions measured were $SO_4^{-2}$, 23% were $NO_{-3}$ and 14% were $Cl^-$. Of the cations measured, 69% were hydrogen.

Therefore, one can conclude that the acidic intensity of rainfall has a relationship to the amount of sulfuric and nitric acids that have been transported by the wind currents from the atmospheric pollution of the urban and industrial areas.

Nine different sampling stations in Pennsylvania and New York determined the pH of precipitation between 1965 and 1973. While some stations have a wider range of pH and more variance than others, the trends of pH in these nine years are about the same.[16]

A comparison of $SO_2$ emissions from electric utilities for eastern United States and ambient sulfate concentrations at Duncan Falls, Ohio reveals that the range of sulfate concentration varied about 400% while the $SO_2$ emissions were constant.[16]

The data collected by Air Quality Services, Inc. and presented by their President, Dr. John O. Frohlizer, at the National Symposium on Acid Rain in 1980 in Pittsburgh, PA shows that there is not a significant change between 1973 and 1977. The 1977 pH values are very slightly higher, averaging 4.15 in 1977 and 4.03 in 1973.[17] After the petroleum shortage problems began in 1973 and the resultant increase in use of coal, one would have expected the pH values to decrease.

These data appear to conflict. The important conclusion is a recognition of the enigma and difficulty in setting meaningful standards for emissions. Also, there is great difficulty in preparing a model for acid precipitation for areas that may be many hundreds of miles away from the emission site.

*Coal Mining Impacts*

Mining activities present many difficult environmental problems. Social and economic disruption of small communities can cause concern as large mining or energy enterprises get underway.

Many coal seams are also natural aquifers. Thus, stripping these seams can disrupt water tables and groundwater patterns. Water quality degradation is a major concern with coal mining activities.

An ERDA report estimates that up to 11,000 miles of stream and 70 reservoirs, primarily in eastern regions, have been adversely affected by coal mining. PH changes can be harmful to fish—hence the effect of acid mine drainage on aquatic life.

Sedimentation from surface mines, erosion of spoil piles created by mining and coal cleaning, and acid mine drainage containing high concentrations of sulfur, iron and maganese have and will continue to cause much environmental concern.

*Summary and Conclusions*

Until alternate renewable sources of energy can be developed, coal and nuclear energy are the primary sources of energy and power generation in the near future. Both large and small scale hydro-electric power generators at both privately and publicly owned dams should be encouraged. Some alternative energy sources such as wind, solar, methane and alcohol systems will contribute in a small but increasing way to the energy needs.

The enigma of how to protect our resources from environmental degradation still exists. Removing solids from water may make the water suitable for some intended use. If the solids are disposed of on land, they may leach back into the water course. If the solids are incinerated, air pollution problems arise from their emissions.

Air, water and solid waste problems are all interrelated. By removing the sulfur from coal, we can reduce the air pollution problems, but then the problem shifts to one of solid waste disposal.

Resource recovery and reuse systems are one solution. Technology must be developed and improved to minimize the amount of waste products that our every day living generates. Reuse will decrease environmental pollution and conserve our natural resources.

The use of fossil fuels such as coal for power generation is necessary, but not without its enigmas. Technology must be developed so that it becomes environmentally compatible to use the fossil fuels for energy production.

## BIBLIOGRAPHY

1. *Historical Statistics of the United States Bureau of the Census,* (Washington, D.C.: U.S. Bureau of Mines, 1974)
2. *The Toledo Edison Company Annual Report 1980,* (Toledo, Ohio: Toledo Edison Co. 1981) p. 3.
3. J. Harte and R.H. Socolow, *Patient Earth* (New York: Holt, Rinehart and Winston Inc., 1971), p. 285.
4. "World Energy Supplies, 1971-1975." U.N. Statistical Papers, Series J. No. 20 (New York: United Nations, 1977).
5. "National Energy Outlook, Executive Summary, 1976." Federal Energy Administration (Washington, D.C.: U.S. Government Printing Office, 1976).

6. "Energy Facts II," Subcommittee on Energy Research, Development and Demonstration, U.S. House of Representatives, 94th Congress, Library of Congress, Serial H. August 1975, p. 44.
7. *A National Plan for Energy Research, Development and Demonstration:* Creating Energy Choices for the Future Washington, D.C.: ERDA Vol. 1, Chapt. 11, U.S. Government Printing Office 1975)
8. *National Air Quality and Emissions Trends Report, 1976* (Research Triangle Park: U.S. EPA, EPA-450/1-77-002, December 1977)
9. L.S. Shannon, P.G. Gorman, and W. Park, *Feasibility of Emissions Standards Based on Particle Size,* Prepared for the Office of Research and Development, United States Environmental Protection Agency, EPA-600/5-74-007, March 1974.
10. *National Air Quality, Monitoring and Emission Trend Report, 1977* (U.S. EPA December 1978)
11. P. Avevitt, U.S. Geological Survey, Bulletin 1136, January 1960
12. C.C. Sun. C.H. Peterson, R.A. Newby, W.G. Vaux and D.L. Keairns, "Disposal of Solid Residue from Fluidized-Bed Combustion: Engineering and Laboratory Studies" Pittsburgh, PA, Westinghouse R & D Center, July 7, 1978.
13. W.H. Griest, M.R. Guerin, "Indentification and Quantification of Polynuclear Organic Matter (POM) on Particulates from a Coal-Fired Power Plant," Oak Ridge, Tenn., Oak Ridge National Laboratory EPRI EA-1092, Project 1057-1, DOE RTS 77-58, Interim Report, June 1979.
14. W.A. Duvel, Jr., W.R. Gallagher, R.G. Knight, C.R. Kolarz and R.J. McLaren, "State-of-the-Art of FGD Sludge Fixation," Beaver, PA. Michael Baker, Jr., Inc., Prepared for the Electrical Power Research Institute, January 1978.
15. Sun, C.C., C.H. Peterson and D.L. Keairns, "Environmental Impact of the Disposal of Processed and Unprocessed FBC Bed Material and Carryover," Westinghouse Research and Development Center. Presented at the 5th International Conference on FBC, February 1979.
16. C.V. Coghill and G.E. Likens, "Acid Precipitation in Northeastern United States," *Water Resources Research,* No. 10, (1974), p. 1133-1137.
17. *Proceedings of the National Symposium on Acid Rain,* Pittsburgh, *September 23-24, 1980,* "The Misconceptions (Myths) About Acid Rain" by John O. Frohiger (Pittsburgh, PA: Greater Pittsburgh Chamber of Commerce and Pennsylvania Society of Professional Engineers, 1980).
18. James E. Hall, "Environmental Impact of Flue Gas Desulfurization (FGD) and Fluidized Bed Combustion (FBC) Solid Waste Disposal" (unpublished M.S. Thesis, Department of Civil Engineering, School of Engineering, University of Pittsburgh, 1979).

19. Gisele A. Williams, "The Enigma of Acid Rain" (unpublished M.S. Thesis, Department of Civil Engineering, School of Engineering, University of Pittsburgh, 1981).

*Chapter Eleven*

# Nuclear Energy and its Role in Society

**Carl Roman, P.E.**
ROMAN RESOURCES &
DEVELOPMENT CORP.
2206 Chew Street
Allentown, Pa. 18104

Carl Roman, P.E., is president of Roman Resources and Development Corp., consultants in engineering and human resource development; president of Lehigh Valley Opportunity Center (LVOC), a non-profit, state-oriented criminal rehabilitation corporation; and founder of, and consultant to, the Technical Advisory Council (TAC) of the Lehigh Valley, a 1980 state (PA) and national award winning and internationally acclaimed engineering/science community service project of the professional engineers society. Carl has written and published TAC's first study, "The Transportation of Nuclear Materials" to inform the public and its legislators about nuclear technology in the aftermath of nearby Three Mile Island. In 1980 he was awarded a nationally competitive fellowship to Georgia Institute of Technology on the effect of technology on American society. He is a member of the Lehigh Valley, Pennsylvania and national professional engineering societies; the Society for Intercultural Education, Training and Research; the International Studies Association and the board of directors of the Program for Female Offenders.

## INTRODUCTION

The word energy has become one of the most used, even abused, words in the world today.

Why?

Because energy no longer means, as dictionaries conventionally define it: the human capacity for vigorous action; or the capacity, according to the physics curriculum, for doing work and overcoming resistance. Energy has become almost synonymous in our Nuclear Age with nuclear or radioactive energy which translates, liberally, into power. Not only electric power; but economic and military power. Power to build, control and even destroy nations. Power to improve the quality of human life through a world-wide industrial revolution, already in progress, that will dwarf the Euro-American industrial revolution of a century ago.

The technological-sociological impact of radioactive energy presents a dichotomy of build and destroy—of life and death. This article will present a factual and informative semi-technical overview of nuclear technology and society including the uses and sources of radioactive energy, basic nuclear technology, the transportation of radioactive materials, societal impact and comparative risks associated with radioactive energy.

## ORIGIN AND SOURCES OF RADIOACTIVE ENERGY

*Origins of Radioactive Energy.* Nuclear energy is a relatively new source of energy that had its origin about 90 years ago when a German physicist, Wilhelm Konrad Roentgen, discovered x-rays and gave birth to nuclear chemistry. Roentgen discovered his x-rays in the nucleus of the fundamental building block of the universe, the atom. The energy potential of the atom was revealed to the world in the form of a military weapon, the atom bomb, that ended WWII in 1945. This event is recognized as the beginning of the Nuclear Age. (1)

*Nuclear Energy: A Continuum of Electric Energy.* Both electrical and radioactive energy have their source in the atom. The practical application of radioactive energy followed electrical energy by about 100 years as science developed better instruments and techniques to probe more deeply into the atom.

Having discovered that the high speed movement of negatively charged electrons orbiting the nucleus of the atom produce electricity (6,280 billion million electrons passing a given point in one second produces one ampere of electric current. See Fig. 1), it seemed scientifically logical to probe the control mechanism of those electrons—the nucleus. With the discovery of the radioactive power in the nucleus, the Nuclear Age arrived. Society was now faced with a challenge similar to the beginnings of electricity—of utilizing this new power without fully understanding it.

FIGURE 1. Electric Energy From the Atom
Source: The Transportation of Nuclear Materials. See ref. #1.

## FUNDAMENTALS OF RADIOACTIVE ENERGY

*Sources of Radioactive Energy.* Radioactive energy, or nuclear radiation, is created when the nuclei of unstable substances absorb or release particles. These radioactive materials can be unstable atoms found in nature, such as radium, thorium and uranium (the amount of thorium and uranium three miles into the earth's crust is estimated at one thousand billion tons); isotopes of common elements like potassium, carbon and hydrogen; or stable atoms made radioactive by bombarding them in a reactor or accelerator.

Other forms of radiation come from nature (soil, water, natural food cycles, cosmic radiation) and from man (x-ray machines, television sets, wrist watches, building materials) power generation, and atomic weapons. (1)

*The Energy Process.* Nuclear energy is created by the decay or disintegration of nuclear material, usually in a chain-like fashion. The end product of every decay chain is a stable non-radioactive material. See Fig. 2.

Uncontrollable by physical or chemical means, the decay process must go through a time period characteristic of the particular material—a major problem in disposing of nuclear wastes. The technical term for measuring the time is "half

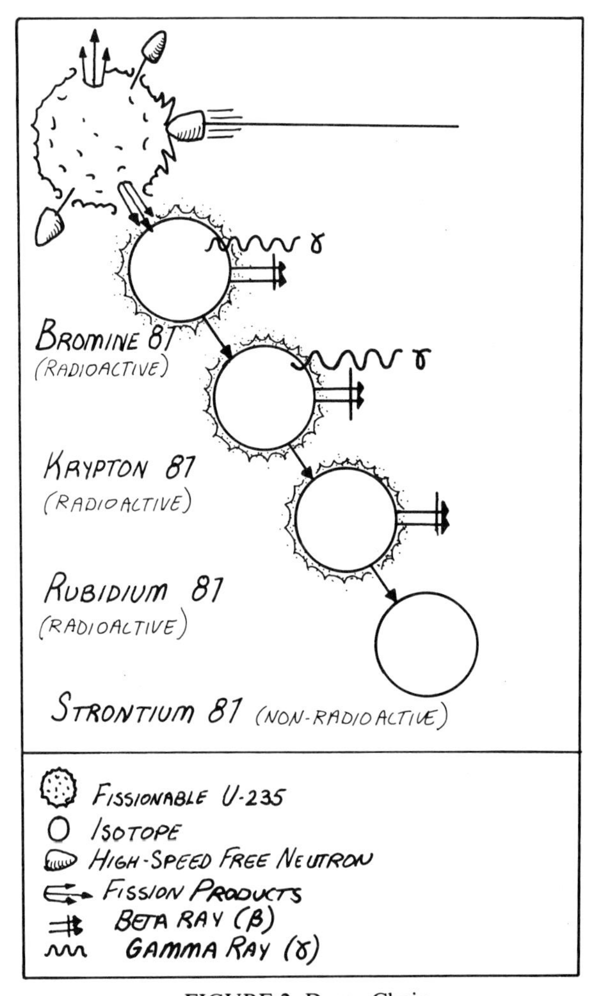


FIGURE 2. Decay Chain
Source: The Transportation of Nuclear Materials. See ref. #1.

life,'' the time to decay the material to one half its original quantity. Uranium-238, an isotope of uranium with a mass number of 238 has a half-life of 4.51 billion years; lead-211 has a half life of 36.1 minutes and polonium-211 has a life time of 0.5 seconds. Figure 2 shows the decay chain of bromine-87, one of the many fission products of uranium-235. (1)

*Energy Components.* Nuclear energy may be released in several forms: particles (positively charged alpha particles, negatively charged beta particles, uncharged neutrons), or electromagnetic rays (gamma rays, x-rays).

The intensity of radiation in free air of all forms of energy from a point source varies according to the Inverse Square Law. The amount of energy each contains and their effect on living and non-living objects varies. Alpha particles upon direct contact will travel only one millimeter into human tissue and can be blocked by a single piece of paper. Gamma rays pass through the body easily (hence their widespread use in medicine) and need large thicknesses of dense material, like lead, to absorb them. (1)

## USES OF RADIOACTIVE ENERGY

*Radioactive energy* has many uses ranging from national defense to food processing. Scientific researchers often use radioactive substances to determine the mechanisms of complicated chemical and biological changes.

*Medicine.* In the medical and health care fields radioisotopes are used for both diagnoses and treatment. Radioisotopes, such as technetium-99, injected into the blood stream permit external measurement of the performance of the organs. Radiation therapy with isotopes of radium-226, gold-198, cobalt-60, iodine-131, cesium-137 and iridium-192 treat cancer patients. Cardiac pacemakers powered by plutonium-238 can operate for 10 years or more before surgical replacement is required.

*Industrial and Commercial.* Radioisotope devices measure and gauge metal thickness. Radiography using cobalt-60 and cesium-137, inspects welds and metal castings for flaws. And radioactive tracer materials are used to find and tag formations of oil and to investigate fluid flow problems in the petroleum industry.

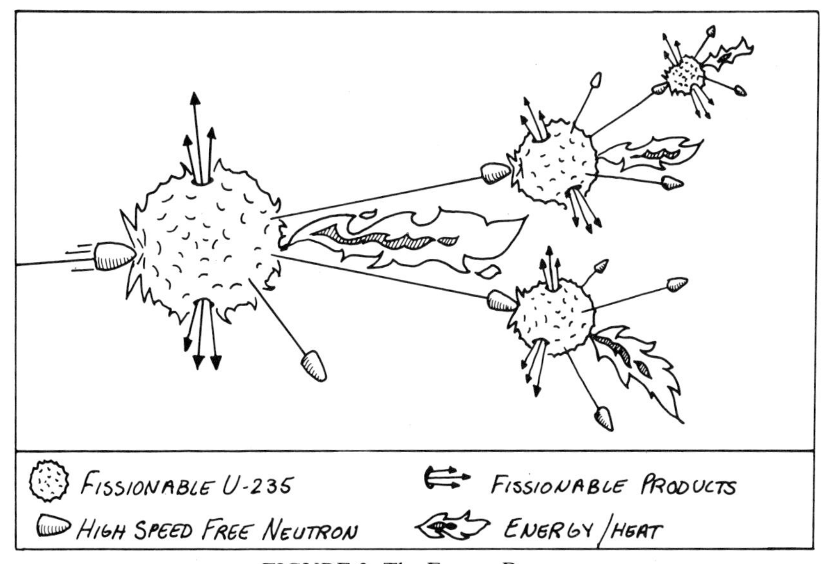


FIGURE 3. The Energy Process
Source: The Transportation of Nuclear Materials. See ref. #1.

Radioisotopes of cobalt-60 and cesium-137 in the growing process radiation industry irradiate materials to destroy bacteria in food and cause chemical changes that improve physical characteristics.

*Military.* Since the development and use of the atom bomb during World War II, this special, high-explosive form of radioactive energy has replaced other forms of explosives for conventional military weapons and is being used as the explosive element in new weapons. Electricity for military and space use is often supplied by nuclear generators. (1)

*Fuel for Electric Power.* One of the more important and most controversial uses of radioactive energy is fuel to supply the heat that creates the steam for electric turbine-generators. Coal and oil and some natural gas have been the traditional fuels. Economics of coal and oil supply (particularly in oil since 1973), concern for the environment and the societal trauma following the Three Mile Island nuclear power plant incident in 1979 have focused attention on abundant nuclear and coal fuels as viable solutions to an energy-needy world.

## SOME PROBLEMS, FALLACIES AND POTENTIAL SOLUTIONS

*Radioactive Fuel Compared to Coal.* Nuclear fission of one ounce of nuclear fuel, uranium-235, in a nuclear reactor produces about the same energy as burning 100 tons of coal—put another way, 38,000 railroad cars of coal equal one railroad car of uranium. Plutonium-239, a radioactive biologically hazardous by-product of uranium-235 fission, has a half-life of 24,400 years. In the still experimental breeder reactor, a kind of nuclear "furnace," plutonium-239 is used as fuel and in the process emerges as a by product that is up to 20% greater in volume than the original fuel. (1) Plutonium presents the dichotomy of providing unlimited fuel while posing a societal risk because of its relative toxicity.

Plutonium-239 plays a central role in reprocessing fuel and producing nuclear weapons; but in consideration of the hazards, the Carter administration had decided to defer indefinitely the commercial processing and recycling of plutonium. The Electric Power Research Institute (7) notes that plutonium is often called the most toxic (or poisonous) substance known to man. Comparison of relative toxicity of chemicals is almost meaningless unless there is some similarity in the comparative techniques. High toxicity quotations on plutonium are based on minimum quantities being injected into the blood stream. A more reasonable intake route through food ingestion would require a minimum dose 30,000 times as large. In experiments at Los Alamos, over 2000 rodents with more than 2 million particulates containing plutonium placed in their lungs showed the same number of caricinomas as a control group with no plutonium.

According to Dr. Harold Agnew, president of General Atomic Co., (10) radiation emitted by coal-fired plants, through uranium in the coal or radiation released through mining the coal, is many times more than radiation released from nuclear-fired plants or mining uranium. Noting that a 1000 megawatt coal-fired plant emits more than a ton of uranium per week, he feels that perhaps the greatest long term danger from coal-fired plants is from the other poisonous materials contained in coal. Plant consumption of 30,000 tons of coal a day could release into the atmosphere each year about 120 tons of lead (the scrubbers take some lead out), 10 tons of mercury and about 30,000 tons of sulphur. And these materials have infinite half lives, as do some stable materials such as arsenic, lead, mercury and thallium (a rat poison). They are here forever. Nuclear power is cheaper by a factor of four over oil and a factor of two over coal, i.e. electricity delivered to the consumer—including fuel, money, maintenance, salaries and construction —would cost four times as much to produce if oil is used and two times as much if coal is used. Moreover, the radiation from a nuclear plant is closely monitored, whereas that from a coal-fired plant is not.

*Public Safety.* The U.S. Department of Transportation (U.S. DOT) has the regulatory responsibility for the safe transport of radioactive materials in interstate or foreign commerce. Postal shipments are regulated by the U.S. Post Office. Shipments not in interstate or foreign commerce are subject to control by state agencies in most cases.

Other major governmental agencies concerned with the safety in handling radioactive materials are: at the federal level, Nuclear Regulatory Commission (NRC) and Atomic Energy Commission; at the Pennsylvania state level, Department of Transportation (Penn DOT) and the Department of Environmental Resources. Subdivisions of these agencies, in addition to establishing safety regulations, become extensively involved in investigating and assessing radioactive material accidents. The derailment of four 16-ton casks of enriched uranium hexafluoride on March 31, 1977, near Rockingham, N.C. brought out 16 different agencies and departments. (1)

Transportation of nuclear materials "is one of the strongest links in the nuclear safety chain" said Dixie Lee Ray, former Atomic Energy Commission chairman. The excellent transportation safety record has been attributed to the attention shippers give to proper packaging, effectiveness of safety standards and regulations and safety guidelines focusing on protection against the consequences of accidents and radiation exposure to people in the vicinity; and shielding radioactive materials to reduce external radiation doses.

Packaging of "low specific activity" material which can, in fact, be highly irradiated is a key area for improving the fine radioactive material transport record. In a letter to the Technical Advisory Council of the Lehigh Valley, October 1979, the Savannah River Operations Office of U.S. DOE stated that federal regulations for packaging are "—too loose—over 60 times in 1978 faulty containers spilled radioactive waste in the trucks of the Tri-State Motor Transit Co., the largest transporter of nuclear waste." (1)

A Sierra Club report on the 1974-1978 record of Tri-State, which hauls about

3000 irradiated shipments a year, said that only 1.27 percent of their shipments resulted in some spillage within the trucks. (1)

In a test by ERDA in 1977, a shipping cask carrying nuclear fuel was rammed into a concrete wall at 60 miles per hour, incurring only minor scratches. In another test a nuclear fuel shipping cask on a truck survived a crash by a locomotive propelled by a rocket at 81.5 miles per hour. The cask received only minor dents and its integrity was not breached by this test. (1)

*Emergency Action.* How are radioactive material accidents handled? Typical emergency action includes lifesaving assistance, measurement of contamination, decontamination of the patient by cleansing and isolation of the patient until decontamination required by the exposure is completed.

The first steps are the same as for any other accident where life is endangered, the environment contaminated, or property is damaged: (1) Persons first on the scene should notify the police, render assistance to any injured persons either by removing them to the closest spot away from fire or explosion danger, and administer first aid. (2) The driver, if capable, and the police should initiate the special emergency procedures applicable to an accident in which nuclear materials are involved. But with radioactive material, special emergency procedures are necessary. "Emergency Action Guidelines for Incidents Involving Radioactive Materials," published by U.S. DOE is perhaps the best of several nearly identical plans of action available. (1)

Federal agencies, mostly DOE and NRC, respond to most radiological accidents. The exception is incidents involving weapons, weapons components or other military nuclear materials. In such cases the Department of Defense would respond. In general, Federal agencies minimize their own involvement and mandate that on-the scene authority remain in the hands of the presiding local agency. One of the objectives in the DOE plan is, "To encourage state and local governments, private industry, and other organizations to develop their own radiological emergency capabilities and plans for coping with radiological incidents." (1)

## SOCIETAL-TECHNOLOGICAL RELATIONSHIPS

*Role of Information and Communication.* To a planet of 8 billion people, many emerging from cultural shells spanning thousand of years, two events in the past decade—the continuing oil crisis that began in 1973, and the Three Mile Island nuclear plant accident in 1979—will greatly influence the course of events in the 21st century. Almost simultaneously, rapid advances in electronic technology have established a world-wide network for the future Global Village. Both events dramatize the bread-and-butter role of energy and the supporting role of nuclear technology in a global technological society.

Developing a societal-technological partnership will depend greatly on an effective world-wide communications network that is capable of exchanging knowledge of nuclear technology clearly and completely. Abraham Lincoln voiced this sentiment during the Civil War when he told our "melting pot" nation that if the people were informed the nation would be saved.

*Building a Partnership.* A successful, technological-societal partnership requires a communication technique that will overcome the emotional and confusing interpretation of technology by the media, and introduce an objective assessment of the threat, or risk, of radioactive materials and other hazardous substances compared to other acceptable risks.

Risk is a fact of daily social life in transportation (cars and airplanes), sports (football and auto racing), hobbies (sky diving, and mountain climbing), and body nutrition (bacteria, alcohol), among others.

Risk, in peoples eyes, is often perceived by less specific and sometimes irrational factors such as: control of the risk situation, catastrophic consequences, and dramatic and emotional effect. Fear of flying is an example of the lack-of-control risk. But auto travel, which has a much worse safety record than flying and cigarette smoking are not considered especially risky; the individual feels in both cases that he has control of these activities and that the consequences of either will not seriously harm him. To statisticians: Risk is defined as the product of how frequently an event occurs and how severe the effects of the occurrence are.

The fear of the presence of radioactive materials is evident in the statisticians perspective of transportation of nuclear materials. In peoples eyes, this risk may be colored by the dramatic and emotional perception by the public of a major "nuclear" accident and the individual's lack of control over such an accident. From a statistician's viewpoint alone, this risk becomes quite complex due to the following rational factors that affect the consequences: (1) was the location in the city or the country? On a heavily traveled crosstown freeway or a rural stretch of turnpike? (2) was the surface dirt, macadam or concrete? Was it dry, wet, icy or snow-covered? (3) what speed was involved? (4) was the accident on a curve or top of a hill, on a grade or on a straight stretch? (5) how many vehicles were involved? (6) what if fire breaks out? How would it affect other factors? The complexity of the situation becomes more apparent when population distribution, geographic conditions and weather factors, which change continuously over the route traveled, are taken into consideration. (1)

Risks associated with various hazardous materials have been made and compared with nuclear risks. Uranium hexafluoride chemically reacts to moisture in the air forming extremely toxic uranyl fluoride and hydrogen fluoride: exposure to concentrations as low as 45 parts per million for several hours can kill a human and its noxious smell is detectable at less than 1 part per million. Shipping cylinders are kept at low pressure and only when fire accompanies an accident is there any significant release of potentially toxic fumes. Radiation hazards in this case are minor compared to the risk of chemical poisoning. (1)

Comparing the minimum doses of other toxic materials, plutonium is about the same as nerve gas (sarin), and much less toxic than many biological agents such as botulism and anthrax—all of which cause death in a few hours. Plutonium in food is roughly hundreds of times less carcinogenic than some of the by-product mycotxins associated with common food molds (aflatoxin $B_1$ is one of the most potent carcinogens known). Plutonium is also much less toxic than lead arsenate, selenium oxide, potassium cyanide, and mercury dichloride—all of which cause a rapid death.

The risk from radioactive material-exposure to radiation may be controlled by (1) minimizing the exposure time, (2) maximizing the distance from the radiation source (the inverse square law for radiation intensity), and (3) shielding the radiation source. Effectiveness of radiation shielding depends upon the radiation energy being absorbed by the shield material. Alpha-rays are the least penetrating, beta-rays next, and gamma-rays are the most penetrating. Alpha and beta radiation can be completely absorbed. Gamma radiation can only be reduced in intensity by thicker absorbers but it cannot be completely absorbed. For absorbing gamma radiation of a specific energy (1MeV), water is 840 times more effective than air as a shield.

Risk of release of radioactivity is the product of the probability of it happening and the consequences if it did happen. In studies made of this risk, a model was developed which allows for variance of many factors defining the total risk as the sum of the risk of each possible accidental release of radioactive material (sabotage was not considered). (1)

Comparisons lend perspective to the safety issue, but they do not determine the acceptability of the risk. The risks which a society is willing to accept must be evaluated rationally, but also accepted emotionally by its members.

## MAINTAINING A PERSPECTIVE

The Nuclear Age is more than nuclear energy; it is an age where man becomes an immediate participant and first hand observer to the best and the worst of history. In his living room he becomes enmeshed in Three Mile Island, lands on the moon, relives the Holocaust, and assists in a heart transplant. These are ongoing events of problems and progress that will be grinding out for decades.

The ability to cope or not to cope with the compounding effect of this historical perspective will provide a major battleground for conflicting philosophies, life styles and perceptions of national lifestyles in the global village. The energy problem may well be the hard core of our copeability.

## REFERENCES

1. "The Transportation of Nuclear Materials," Carl Roman, P.E., The Technical Advisory Council of the Lehigh Valley, Feb. 1, 1980.

2. "Radiation Risks for Nuclear Workers," Atomic Industrial Forum, Inc., Washington, D.C.
3. "The Nuclear Controversy," Ralph E. Lapp, Fact Systems, April 1975.
4. "Socially Responsible Energy Futures," Chauncey Starr, P.E., Professional Engineer, Oct. 1979.
5. "The Way to Save Nuclear Power," R.A. Brightsen, Professional Engineer, Oct. 1979.
6. "Necessity of Fission Power," H.A. Boethe, Scientific American.
7. "Nuclear Safety and Protection," Chauncey Starr, P.E., Electric Power Research Institute, April 1975.
8. "Nuclear Issues and the Public," L.R. Wallis, General Electric, Sept. 10, 1976.
9. "Nuclear Reactor Safety," Atomic Industrial Forum, Inc., Washington, D.C., Sept. 1976.
10. "Nuclear Power: A Necessity," H.M. Agnew, General Atomic Co.

*Chapter Twelve*

# The Bituminous Coal Industry of Pennsylvania Localization and Trends

**E. Willard Miller, Ph.D.**
Professor of Geography (Emeritus)
Department of Geography
302 Walker Building
PENNSYLVANIA STATE UNIVERSITY
University Park, Pennsylvania 16802

Dr. E. Willard Miller is Professor of Geography and Associate Dean for Resident Instruction Emeritus, in the College of Earth and Mineral Sciences of The Pennsylvania State University. He received his B.S. degree from Clarion State College and Ph.D. from Ohio State. In 1975 he received a citation from the Governor of Pennsylvania for service to the commonwealth. In 1948-47 he was President of the American Society for Professional Geographers. He is an authority on minerals and manufacturing having written more than 100 articles and 16 books on these fields. Dr. Miller was president of the Pennsylvania Academy of Science 1966-68 and for the past five years has written the Earth Scientist's Corner for the Academy Newsletter.

Bituminous coal mining began in western Pennsylvania with the first settlements in the 18th century. During the first three-quarters of the 19th century production was modest reaching only 1,305,932 tons in 1877. With the expansion of the iron and steel industry in the Pittsburgh region production expanded to 79,318,362 tons in 1900 and continued to grow rapidly to an initial peak of 177,217,294 tons in 1918. In the 1920's production was erratic, but gradually declined to 142,351,359 tons in 1930. With the economic depression of the 1930's, production plummeted to a low of 74,166,485 tons. During the World War II period bituminous output once again increased reaching a high of 144,761,964 tons in 1947. After this date production declined steadily to a low of 63,171,313 tons in 1961, the lowest output since 1898. However, since 1961 production has steadily grown to 80,491,000 tons in 1970 and 89,166,000 tons in 1979. The purpose of this paper is to determine the spatial changes in Pennsylvania's bituminous industry during the present growth period beginning in 1961 and to interpret and analyze some of the modern day trends in coal production.

## METHODOLOGY

In order to determine the spatial patterns of coal production in 1961 and 1979 a central tendency technique is utilized. Although the central tendency measure reveals spatial patterns, it does not reveal the dynamics of change. The magnitude of the changes, either gain or loss, are determined by using net shift analysis.

## PENNSYLVANIA'S PLACE IN THE NATION

### Localization in 1961

Coal production in the United States, just as in Pennsylvania, reached a post World War II low in 1961 when only 402,977,000 tons were produced. At this date coal production in the 27 producing states varied from a low of about 100,000 tons in Georgia to a high of 113,071,000 tons in West Virginia. Pennsylvania was the third largest producer.

It was evident that the states had experienced differential rates of growth. Further, it was apparent that the data distributions were skewed and exhibited log normal characteristics. An empirical approximation of this characteristic is maintained by dividing the above mean observations into three equal groups, and those below the mean into three equal groups (Figure 1).

To provide a graphic presentation for comparative purposes with the maps, a chart is included in which the rank ordered states are plotted against value, and the resulting curve is shown on the graph. To the left of the graph is the range of values associated with the three groups of states above the mean. Directly below the graph a horizontal bar shows the number of states in the three categories above the mean. Thus, it can be quickly determined how many states are in each category and the range of values included in each category.

In 1961 there were seven states located above the mean and 20 states below the mean. The seven states above the mean had a production of 362,396,000 tons of coal, or 89.9 percent of the total. These seven states in order of production were West Virginia, Kentucky, Pennsylvania, Illinois, Ohio, Virginia, and Indiana (Figure 1A). They were thus all located in the Appalachian or Eastern Interior Coal Provinces. The 9 states of the Appalachian Coal Province—Alabama, Georgia, Eastern Kentucky, Maryland, Ohio, Pennsylvania, Tennessee, Virginia and West Virginia—produced 290,429,000 tons or 72.07 percent of the total. Pennsylvania alone produced 15.54 percent of the bituminous coal of the nation in 1961. The eight western coal producing states — Arizona, Colorado, Montana, New Mexico, North Dakota, Utah, Washington, and Wyoming—had an output of only 14,995,000 tons, or 3.7 percent of the national total.

## Localization in 1979

By 1979 coal production in the United States had risen to 767,856,000 tons. The United States mean was thus 28,439,000 tons. The eight states—Kentucky, West Virginia, Pennsylvania, Wyoming, Illinois, Ohio, Virginia and Montana—located above the mean had an output of 590,815,000 tons of bituminous and lignite coal, or 76.9 percent of the total (Figure 1B). Six of the states appeared above the mean in both 1961 and 1979, but most significant is the appearance of Wyoming and Montana as new major producing states in 1979. Pennsylvania remains the third largest coal producing state. Its relative position has however declined with an outout of only 11.9 percent of the national total. Most significant has been the rise of the eight western states where production in 1979 totaled 174,325,000 tons, or 22.7 percent of the national total.

## Net Shift 1961-1979

A basic question now is, with a national geographical pattern that exhibited considerable spatial instability between 1961 and 1979, how has the position of Pennsylvania changed in the bituminous coal industry?

The analysis is based on actual coal production for each state for 1961 compared with a theoretical figure showing what the coal production of that state would have been if it had grown at the same rate as the United States between 1961 and 1979. Between 1961 and 1979 coal production increased by 90.54 percent in the United States. If coal production grew more rapidly in a state than in the United States, it is said to have experienced a "comparative gain," if it grew less rapidly, it is said to have experienced a "comparative loss." A comparison of the actual 1979 with the theoretical 1979 figures for a state gives the magnitude of the growth or decline.

Of the 27 coal producing states, 15 experienced a comparative growth and 12 a

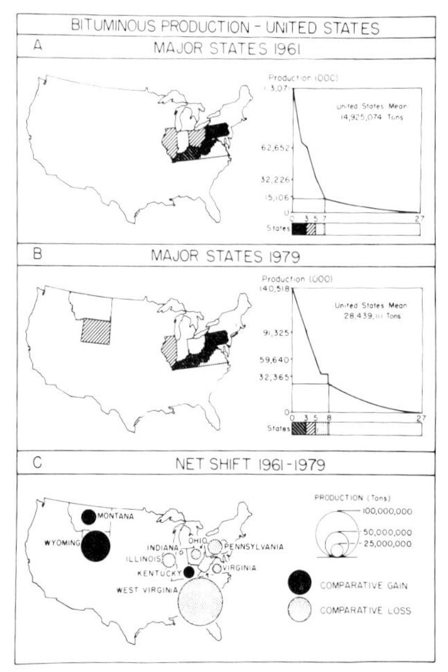


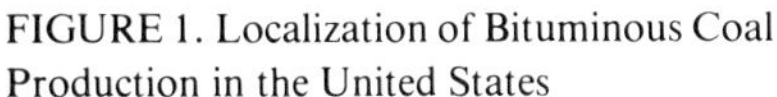
FIGURE 1. Localization of Bituminous Coal Production in the United States

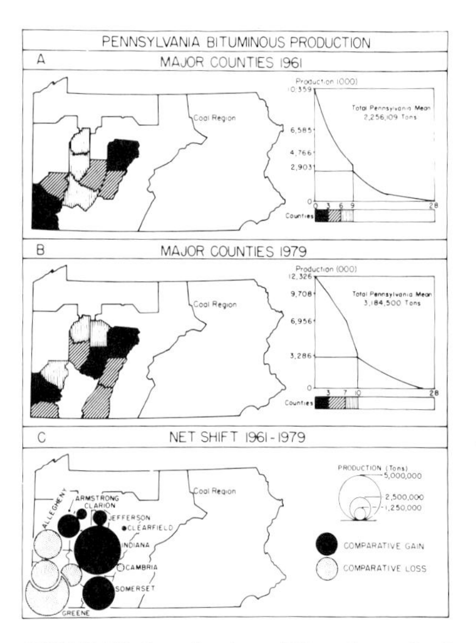


FIGURE 2. Localization of Bituminous Coal Production in Pennsylvania

comparative decline (Figure 1C). However, of the nine states above the mean in 1961 and/or 1979 six experienced a comparative loss ranging from 1,515,000 tons in Indiana to 103,705,000 tons in West Virginia. Pennsylvania had a comparative loss of 28,052,000 tons of bituminous coal. Of the seven states in the Appalachian and Eastern Interior Coal Provinces six states experienced comparative declines, only Kentucky had a comparative gain. Wyoming had the largest comparative gain of 65,385,000 tons, followed by Montana with 21,659,000 tons. The comparative growth was thus primarily outside the traditional areas of coal mining. The eight western states experienced a comparative growth of 134,288,000 tons. If these eight states had grown at the national rate of production their output would have been 28,571,000 tons, instead their actual production was 174,325,000 tons. In contrast if the 9 states of the Appalachian Province would have had a production increase at the national rate of growth, they would have produced 553,383,000 tons. Instead the actual Appalachian Province production was 426,830,000 tons.

## LOCALIZATION OF COAL PRODUCTION IN PENNSYLVANIA

In Pennsylvania while total bituminous coal production increased from 63,171,000 to 89,166,000 tons between 1961 and 1979, the underground production increased slightly from 41,958,000 to 43,350,000 tons—an increase of only 3.31 percent, while surface mining output increased from 20,744,000 to 45,116,000 tons, an increase of 117.48 percent. As a consequence, not only the localization of total production, but also of surface and deep production, must be considered.

### Total Production

Localization in 1961-1979. In 1961 production in the 28 bituminous coal counties of Pennsylvania totaled 63,171,000 tons. The Pennsylvania mean was 2,256,107 tons. Production, however, varied from 6,000 tons in McKean County to 10,359,000 tons in Washington County. There were 9 counties located above the mean and 19 counties below the mean (Figure 2A). The 9 counties above the mean had a total production of 51,571,000 tons, or 81.6 percent of the total. Further, the top three counties—Washington, Greene and Clearfield—produced 26,520,000 tons, or 41.9 percent of the state's total.

By 1979 production had risen to 89,166,000 tons and the Pennsylvania mean was 3,184,500 tons. There were 10 counties above the mean with a production of 76,125,000 tons, or 85.3 percent of the total. Eight of the same counties were above the mean in 1961 and 1979 (Figure 20). Somerset and Jefferson were new counties in 1979, and Westmoreland County had disappeared from above the mean.

Net Shift 1961-1979. Although the geographical pattern remained quite stable between 1961 and 1979, the concentration of production changed significantly as revealed by net shift analysis. For a county to experience a comparative gain, its 1979 production had to be more than 141.15 percent of that of 1961. Of the 11 counties above the mean in 1961 and/or 1979, six counties experienced comparative gains from 414,000 tons to 5,440,000 tons while five counties experienced comparative losses ranging from 667,000 tons to 4,701,000 tons (Figure 2C). The largest comparative loses were experienced in the counties that were dominated by underground mining and the largest comparative gains occurred in the counties where surface mining dominated.

### Deep Mine Production

Localization 1961-1979. In 1961 underground production totaled 41,958,000 tons from 26 counties. The underground mean was 1,613,769 tons. Six counties were above the mean with an output of 36,286,000 tons, or 86.48 percent of the total (Figure 3A). The 2 counties in the top one-third above the mean—Greene and Washington—produced 18,960,000 tons, or 45.18 percent of the total underground coal output.

By 1979 deep coal production had increased to 43,350,000 tons. There were 7 counties above the mean of 1,667,307 tons (Figure 3B). These counties with an output of 40,697,000 tons had 93.88 percent of total underground production.

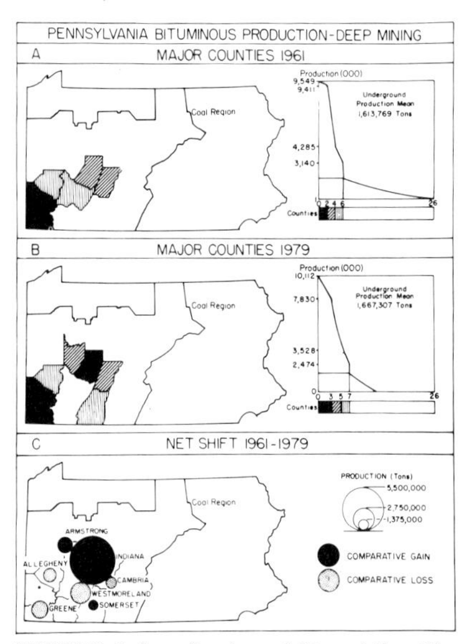


FIGURE 3. Localization of Deep Mine Bituminous Coal Production in Pennsylvania

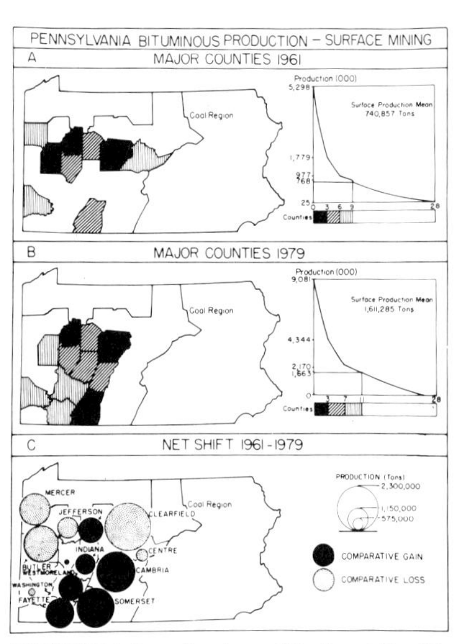


FIGURE 4. Localization of Surface Mine Bituminous Coal Production in Pennsylvania

Further the top 3 counties above the mean—Indiana, Washington and Greene—produced 27,623,000 tons, or 63.72 percent of the total underground production. Of the 8 counties above the mean in 1961 and/or 1979, 5 appear in both years while one of the 1961 counties—Westmoreland—disappeared and Armstrong and Somerset counties are added. However, while 26 counties had underground mining in 1961 only 13 counties had underground mines in 1979. Two major regions of underground mining are evident in 1979—the one is in southwestern Pennsylvania consisting of Allegheny, Washington and Greene counties while the other extends in an arc consisting of the four counties of Armstrong, Indiana, Cambria and Somerset.

Net Shift 1961-1979. For a county to experience a comparative growth it had to have an underground production 3.31 percent greater in 1979 than in 1961. The largest producing counties experienced differential rates of growth from 1961 to 1979 (Figure 3C). Five of the 1961 leading counties—Allegheny, Cambria, Greene, Washington and Westmoreland—experienced comparative losses of 1,748,000, 1,408,000, 2,035,000, 41,000, and 2,654,000 tons, respectively. Three of the 7 counties above the mean in 1979 experienced comparative gains ranging from 1,374,000 tons in Somerset to 5,686,000 tons in Indiana County. However, of the 26 counties producing underground coal in 1961 and/or 1979, 22 experienced comparative losses and only 4 had comparative gains. Indiana had the largest comparative gain primarily due to the mine-mouth electric power plants constructed in that county.

### Surface Mine Production

Localization 1961-1979. Surface mining was widely distributed in both 1961 and 1979. In 1961, when 25 counties had surface mining, there were 9 counties above the mean of 740,857 tons (Figure 4A). These 9 counties with a production of 16,085,000 tons had 77.5 percent of the total. The 3 top counties—Clearfield, Clarion and Butler—had 48.28 percent of total output. Strip mining output in 1961 was localized in a line of 7 counties extending from Mercer County on the west to Centre County on the east with two outliers in Somerset and Washington counties to southwestern Pennsylvania.

By 1979 the surface mean had risen to 1,611,285 tons. As a response to a modest decentralization of production, 11 counties were located above the mean (Figure 4B). These 11 counties dominated production with an output of 38,319,000 tons, or 84.9 percent of the total. The top 3 counties above the mean —Clearfield, Clarion and Somerset—produced 18,817,000 tons, or 41.7 percent of total surface production. Seven of the counties were above the mean in both 1961 and 1979. Mercer and Centre counties disappeared by 1979 as counties above the mean, and Indiana, Fayette and Westmoreland counties were new counties. In 1979 a compact block of counties, extending from Butler, Clarion, Jefferson and Clearfield on the north to Fayette and Somerset on the southern border of the state, was the core of surface mining in Pennsylvania.

Net Shift 1961-1979. Although the geographical pattern of surface mining exhibited more stability than that of underground mining, the counties experienced differential rates of growth. For a county to experience a comparative growth, its production in 1979 had to be 117.48 percent of that of 1961. Of the 28 counties, 11 counties experienced comparative gains ranging from 121,000 tons in Elk County to 2,362,000 tons in Cambria County (Figure 4C). Seventeen counties experienced comparative losses ranging from 54,000 tons in Fulton County to 2,441,000 tons in Clearfield County. Of the 13 counties above the mean in 1961 and/or in 1979, 7 counties experienced comparative gains and 6 comparative losses of production. Of the 6 counties with comparative losses, 5 — Mercer, Butler, Clarion, Clearfield and Centre — were on the northern border of the coal region.

## STRUCTURE OF BITUMINOUS COAL MINING IN PENNSYLVANIA

The bituminous coal industry of Pennsylvania is dynamic reflecting changing economic and political conditions. In its history it has experienced periods of growth as well as decline. The industry is undergoing a number of changes in this present period of growth.

### Reserves and Quality of Bituminous Coal

The demonstrated total bituminous coal reserves of Pennsylvania are 31,000.6

million tons. The coal reserves of the Commonwealth are great, exceeded only by Illinois and West Virginia. The reserves to be recovered by underground mining far exceed those for surface mining. The underground reserves total 29,819.2 million tons while those recoverable by surface mining total only 1,181.4 million tons. The future of the bituminous industry will thus depend on the ability to mine the underground coal deposits.

One of the major problems in utilizing the bituminous coal of Pennsylvania is its high sulfur content. Of the underground deposits only 24.0 percent have 1.0 percent or less sulfur while 54.3 percent have from 1.1 to 3.0 percent sulfur and 11.9 percent have more than 3.0 percent sulfur. The sulfur content is undetermined for 9.6 percent of the underground deposits. Of the surface deposits only 23.6 percent have under 1.0 percent or less sulfur and 12.2 percent have over 3.0 percent sulfur. The sulfur content is undetermined for 9.5 percent of the surface deposits.

Most of the coal mined in Pennsylvania ranges in sulfur content from 1.0 to 2.3 percent. These high sulfur coals do not have a market unless the sulfur is removed. There is an urgent need to improve the processes to remove sulfur from the coals at the lowest possible cost.

### Trends in Underground and Surface Mining

The trends in coal production since 1961 in Pennsylvania are significantly different for underground and surface mining. In underground mining production rose from 41,958,000 tons in 1961 to a modern peak of 56,055,000 in 1969. Since then deep mine production declined sharply to a low of 33,790,000 in 1978, but recovered in 1979 to an output of 43,350,000 tons. In contrast surface mine production was nearly stable from 1961 to 1969 with an output in 1961 of 20,744,000 tons and only 21,652,000 tons in 1969. After 1969 strip mine production rose consistently to a high of 46,552,000 tons in 1978.

Because of increase in total production, the relative importance of underground mining has declined significantly while strip mining has grown in importance. In 1961 underground mines produced two-thirds of the bituminous coal while surface mines had one-third of the total output. This proportion did not change significantly until the early 1970's when strip mine production grew rapidly in relation to total output. By 1979 underground mines were producing 49.4 percent of the bituminous coal of Pennsylvania and strip mines 50.6 percent. Because annual strip mine production varies more than deep mine production, the relative importance of these two types of mining can vary significantly from one year to the next.

### Productivity

The average tons per man per day, when surface and deep mine productivity are combined, has changed little since 1961. In 1961 the average tons per man per day was 11.69 ton and in 1976 it was 11.02 tons. This figure, however, does not reveal significant changes in the average tons per man per day in the deep mines. In the underground mines the productivity per man per day rose from 10.0 tons in 1961 to a high of 12.2 tons in 1970. Since then there has been a drastic decline in productivity to only 8.1 tons per man per day in 1976. The decline in productivity of underground mines is largely a response to the health and safety regulations passed by the Federal government in the early 1970's.

Productivity per man in strip mines has exhibited the same trends as in the underground mines. In 1961 the average tons per man per day in strip mines was 17.37 tons. This output increased to 24.18 tons in 1972 but fell to 17.64 tons in 1976. The decline in output per man reflects the greater depth of coal mined by strip operations and thus the greater overburden to be removed before the coal can be mined. Productivity is thus lowered.

Employment Trends

Although production of bituminous coal reached its lowest level in 1961, employment continued to decline from 29,633 mines in 1961 to a low of 22,969 in 1969. Since then employment has risen steadily to a high of 39,320 in 1978. In 1979 there was a small decline to 37,051.

Employment in underground and surface mining followed essentially the same trends as total employment. In underground mining employment declined from 22,951 in 1961 to a low of 17,771 in 1970. There was then a rapid increase to 24,719 employees in 1973. Since then employment has been essentially stable and in 1979 employment totaled 23,535. Strip mining employment also declined from 6,601 in 1961 to a low of 4,132 in 1969. After a period of stability to 1972, employment rose from 4,553 in 1972 to 10,139 in 1973. In 1974 it declined to 6,416, but since then it rose to a peak in 1978 of 11,117. In 1979 employment was 10,189.

Employment trends reflect productivity in the mines. In the 1960's when output in the underground mines were rising employment declined because of increasing productivity. In the 1970's production in the underground mines has declined but employment has risen reflecting the decline in productivity per man.

A similar situation occurred in employment in strip mining. During the 1960's production was nearly stable but employment declined due to increasing output per man per day. In the 1970's while productivity has declined, production has also risen so there has been a rapid increase in employment in surface mining. However, employment in strip mining fluctuates considerably depending upon the demand for coal at a given time. The place of coal in the energy economy of the nation has varied depending upon the energy crisis of the day. Because strip mines are easily shut-down and easily opened-up there have been rapid changes in employment.

Number and Size of Mines

The total number of bituminous mines has increased only modestly since 1961 when there were 1,763 underground and surface mines in operation. In 1978 the total number was 2,066 and in 1979 it had declined to 1,882. The significant change is the decline in the number of deep mines and the increase in surface mines. There has been a steady decline of deep mines from 1,130 in 1961 to a low of only 144 in 1979. The small underground mines has experienced the greatest decline in numbers. In contrast the large deep mine has grown in importance. In 1975 there were 34 deep mines with production of 500,000 or more tons annually. This compares with only 7 strip mines with production of 500,000 tons or more annually. The decrease in the number of deep mines is directly related to the increasing cost of safety measures. The economies of scale are sufficient in the larger mines to bear the higher costs.

The trends in the number of strip mines exhibit a different pattern from that of the underground mines. In 1961 there were 633 strip mines, but this number declined to a low of 542 in 1944. The number of strip mines rose slowly until 1973 when 903 were operating. Since then the number rose rapidly to a peak of 1,864 in 1978, but declined slightly to 1,738 in 1979. A large number of the strip mines have a small production. In 1975 of a total of 661 mines surveyed, 461 had a production of less than 50,000 tons annually. The largest strip mine operations have diesel drag lines that can move thousands of tons of rock per day, but there are many small operations mining coal seams that have little overburden. The life of a strip mine can then vary from a few months to many years.

## Coal Mine Technology

The modernization of the bituminous coal industry of Pennsylvania compares favorably with other states. In 1975 of 44,631,000 tons of coal produced in deep mines 40,511,000 tons were mined by continuous mining machines, 3,081,000 tons by longwall machines, and 1,036,000 by cutting machines. The mining of coal by hand operations has essentially ceased in Pennsylvania. Of the 705 roof bolting drills in the coal region, 379 were rotary, 173 rotary percussion, and 153 were percussion. A number of methods are used to transport the coal out of the mine. These include locomotives, rubber-tired vehicles and haulage conveyors. Conveyor belts are increasing in importance and totaled 798 units in 1975 with 320.1 miles of belts. Over 97 percent of the coal is loaded into haulage units by the continuous-mining and longwall machines.

In the strip mining industry the earth-moving equipment has gradually become larger in order to reach the deeper surface seams. Overburden of more than 75 feet is now removed to reach the coal seam. In 1975 of 960 power shovels and dragline excavators, 910 were diesel, 27 diesel-electric, 14 electric and 9 gasoline. This equipment varied greatly in the cubic yard capacity of the dipper or bucket. Of the 960 shovels and draglines 654 had a capacity of less than 6 cubic yards, 296 had a capacity of 6 to 18 cubic yards and 10 had a capacity of 16 to 50 cubic yards. All rock drills utilized in the stripping process are either horizontal or vertical power drills.

The cleaning of the bituminous coal from the Pennsylvania fields has lagged. In 1975 when the underground mines produced 44,631,000 tons of coal 33,996,000 tons were mechanically cleaned, or 76.17 percent of the total. However of the 39,105,000 tons produced in strip mines only 8,538,000 tons, or 21.83 percent, were mechanically cleaned. The dispersal of the strip mines and the short life of the mines has discouraged the costly investment in mechanical cleaning equipment.

Method of Shipment

The method of shipment of coal to market has experienced considerable change since 1961. In 1961, 55.8 percent of the bituminous coal was marketed by rail, 26.9 percent by water and 17.2 percent by truck. The percentage marketed by rail has declined steadily to a low of 35.5 percent in 1977. In 1979, 39.0 percent was marketed by rail. The coal marketed by water increased slightly to a high of 32.1 percent in 1969, but since then declined to a low of only 14.9 percent in 1978. In contrast truck haulage has become increasingly important. The growth was small each year until the 1970's when truck haulage of coal increased from 25.6 percent of the total in 1971 to a high of 48.1 percent in 1978.

The increase in truck transportation is directly related to the growing importance of surface mining. The strip mines are widely scattered and railroad and water transportation is not available. Further, there has been a decline in the railroad network in the Appalachian coal region. Because many of the strip mines are small they cannot justify the cost of extending a rail line to them. Trucks are thus best adaptable to small-scale production that extends over a limited time span.

Utilization

The markets for Pennsylvania bituminous coal experienced certain shifts in the 1970's. Electric utilities have grown in importance taking 36.0 percent of the coal in 1970 but increasing to 52.8 percent in 1975. The consumption of coal at mine-mouth generating plants in Armstrong and Indiana counties has increased from 5,847,000 tons in 1970 to 7,237,000 tons in 1975. The average sulfur content of the coal used by electric power generating plants in 1970 was 2.3 percent but had declined to between 2.1 and 2.2 percent in 1975.

In contrast to the increased consumption of coal in the electric utilities the consumption has declined relatively in coke and gas plants from 39.27 percent in 1970 to 25.26 percent in 1975. Industrial, retail and other uses have also declined slightly from 17.93 percent to 12.21 percent of the total consumed. The export of Pennsylvania coal has grown in total volume from 2,897,000 tons in 1970 to 5,059,000 tons in 1975, although there was a small relative decline from 6.25 percent of the total to 5.66 percent. Lack of port facilities to handle shipment has limited the export potential.

## LEGISLATION AFFECTING THE BITUMINOUS COAL INDUSTRY

There are 4 major areas of legislation that have been enacted at the state and/or federal levels that affect the mining of coal. These are laws controlling strip mine reclamation, acid mine drainage, clean air and the health and safety of the miner.

### Strip Mining Reclamation

Although the ravages of the land due to strip mining have long been evident in Pennsylvania, effective controls of strip mining evolved slowly. In the early days of strip mining there was the prevailing attitude that the land in Appalachian Pennsylvania had little value and therefore did not justify the cost of reclamation. This attitude was fostered by a number of considerations. Because the abandonment of farmland began in western Pennsylvania about 1890, long before strip mining was practiced, there was little public or private interest in reclaiming spoil banks on land that had little or no economic productivity. As land became unproductive the practice developed for farmers to have their land strip mined in order to secure a windfall profit, frequently at the time the farmer retired. Although western Pennsylvania could support a forest covering there was little interest in recovering the land by reforestation. The strip-mined areas were dispersed and relatively small so that a solid forest stand was difficult to achieve and interest by major lumber companies was minimal. Finally, the rugged coal mining regions of western Pennsylvania were isolated so that the tourist industry was only modestly developed. Consequently, the esthetic value of the region was little appreciated by outsiders.

The first legislation to develop strip mine reclamation in Pennsylvania was enacted in 1945. The Pennsylvania Bituminous Coal Open Pit Mining Conservation law, the most comprehensive of its day, required that each mining company had to deposit a filing fee of $100 for each stripping operation, and post a bond of $300 per acre to be stripped with a minimum of $3,000. Liability under the bond was for the duration of open pit mining at each operation, and for a period of five years thereafter. This initial act also required each strip-mine operator to cover the exposed face of the unmined cost within one year after completion of mining, and to level and round-off the spoil banks sufficiently to permit the planting of trees, shrubs, or grasses. The slope of the leveled area was not to exceed 45 degrees. After the leveling was completed, the miner was to plant the stripped area to the specifications of the State Department of Forests and Waters. If the operator failed to comply with these regulations, he forfeited all or part of the posted bond.

Although the initial law appeared to provide the necessary regulations for the reclamation of strip-mined land, it was essentially ineffective for a number of reasons. First, the forfeiture of the required bond was not a sufficient penalty to encourage land reclamation. It was estimated that land reclamatiion in Pennsylvania cost two to six times the posted bond. As a result, at least 80 percent of the bonds were forfeited by the mining companies. Secondly, even when the bond was forfeited there was no legal means of preventing the mining company from securing another concession to strip-mine a new area. As a consequence, violators of the law continued to strip-mine areas. Third, the reclamation of stripped land presented difficulties not previously encountered in the revegetation of an area. The average coal mining company had no real interest in a reclamation program and used methods for revegetation that were essentially ineffective. As a result, on many strip mined areas where the spoil banks were leveled and planted in trees, from 60 to 100 percent of the trees died within a year after planting. Finally, state inspection of strip-mine sites was non-existent or at best superficial in many areas. As a consequence the state has a heritage of strip-mined areas from the past that have not been reclaimed.

In order to make the Pennsylvania Surface Mining Conservation and Reclamation Act effective it was revised in 1963, 1968, 1971, 1972, and 1974. The present act is recognized as a model law for the control of strip mining and was used to formulate the national strip mining law of 1977. The present Pennsylvania law requires that the mining company secure an operator's license, mining permit, post a bond, and provide a reclamation plan before strip-mining can begin. The reclamation plan must include the following: (1) a statement of the best use to which the land was put prior to the commencement of surface mining, (2) the use which is proposed to be made of the land after reclamation, (3) where conditions permit, the manner in which topsoil and subsoil will be conserved and restored and, if these conditions cannot be met, what alternative procedures are proposed, (4) where the proposed land use so requires, the manner in which compaction of the soil and fill will be accomplished, (5) a complete planting program providing for the planting of trees, grasses, legumes, or shrubs, or a combination approved by the state Department of Environmental Resources, (6) a detailed timetable for the accomplishment of each step in the reclamation plan, and the operator's estimate of costs, (7) the written consent of the landowner, which allows access to the land for five years after mining ceases, in order to restore the land, (8) the application for a license or renewal shall be accompanied by a certificate of insurance certifying that the applicant has in force a liability insurance policy of not less than $100,000, (9) the manner in which the operator plans to direct surface water from draining into the pit, (10) no approval shall be granted unless the plan provides for a practical method of avoiding acid mine drainage and preventing avoidable siltation or other stream pollution, and (11) the application of health and safety rules necessary for the safety of the mines and public welfare.

The value of Pennsylvania's strip-mine reclamation law lies primarily in the type of enforcement that is practiced. Although the law has been in effect for many years two viewpoints are still expressed. The one viewpoint by conservationists is that if the law is enforced strip mining will leave no scars on the landscape and the land will be economically productive. But the conservationist believes that too frequently the laws are disregarded. The other viewpoint by many mining companies is that the Department of Environmental Resources has been unfair and inconsistent in applying the state's Clean Stream Law and Surface Mining Act. Although much of the strip-mined land is reclaimed, there is also little doubt that reclamation is being disregarded in many areas.

There are a number of reasons for lack of reclaiming the stripped land. In 1981 the state had only 51 inspectors to cover the entire coal region. It was found that some mines had not been inspected in the past year. The lack of proper inspection has resulted in mining and reclamation violations such as not replacing top soil properly, silting from the mining operations and lack of control of acid mine drainage. Mining has become so intense in some areas, for example, in the spring of 1981 there were 20 active mines along a five mile stretch of Popular Run, a tributary of Indian Creek in Fayette County, so that problems of pollution are essentially inevitable. State law forbids the Department of Environmental Resources to renew the annual mining license of a company with environmental violations, but this law has been violated to some extent in recent years.

A few years ago it was assumed that coal would immediately play a greater role in the nation's energy economy. As a response the number of strip mining companies increased rapidly. However, the market for coal has developed very gradually and a large number of companies have been forced into bankruptcy. As a consequence, a large amount of strip-mined land remains unreclaimed.

Bonds posted by coal operators to cover the state's cost of restoring abandoned strip-mined areas are typically too low to pay for the work. As a consequence, DER Secretary, Clifford L. Jones, has raised the minimum bond level from $1,000 an acre to $2,000 an acre in 1980, and in January 1981 to $4,000 an acre. Many conservationists consider this amount is still too low and have indicated that they will attempt to raise the bond level to $6,000 per acre. Secretary Jones has proposed a special reclamation fund to be financed by fees levelled on the coal operators. In contrast, the coal operators indicate that a bond of $2,000 would be adequate if the Department of Environmental Resources did a better job of inspection and enforcement. Many operators have difficulty obtaining the higher bonds from surety companies.

### Acid Mine Drainage

Acid mine drainage is one of the oldest problems in the coal region. It began with the first mining operation and it is estimated that 80 percent of the acid mine drainage pollution today comes from abandoned mines. Within the bituminous region acid mine drainage is coincident with the mining of Allegheny, Conemaugh and Monongahela group coals. The quality and quantity of mine drainage pollutants produced from a mining operation depends upon such factors as to whether the mine is active or inactive, hydrologic, geologic and topographic features of the surrounding terrain, the type of mining method employed and the presence of air, water and iron sulfide minerals.

In surface mines the discharge is often intermittent, generally occurring during and after periods of precipitation. Runoff in stripped areas may find its way directly to surface streams, or may be trapped in pools for a period and only reach a stream during high water periods. Those concentrated "slugs" of mine drainage pollution may damage aquatic life so that it will be permanently harmed. Between flush-out periods, the pools in stripped areas often drain slowly into the backfill to emerge as mine drainage seepages downslope from the mining operations.

The removal of mining refuse materials from mines and coal preparation plants is one of the most difficult problems associated with mining. Refuse piles are a major source of acid drainage. It is also a frequent source of the fine coal and silt pollution common to streams in the coal region.

The attempts to control acid mine drainage were limited until the Federal Water Pollution Control Act of 1972, and as amended in 1977, it is now known as the Clean Water Act. The basic objective of this act, "is to restore and maintain the chemical, physical, and biological integrity of the nation's waters." The major goal is to eliminate the discharge of pollutants into the navigable waters by 1985. It is questionable that the totality of this goal can be achieved in so short a period of time. The technology is available to control acid mine drainage either by the prevention of acid formation or by corrective measures after its formation. However, the processes are costly and modern estimates indicate that several billion dollars are required to clean-up the western Pennsylvania streams alone.

## Clean Air Emission Standards

On December 23, 1971 the federal Environmental Protection Agency issued air pollution control regulations for the emission of sulfur dioxide ($SO_2$) particulates (fly ash) and oxides of nitrogen ($NO_x$). Of these, the emission of sulfur dioxides from coal-fired plants using high sulfur coals, such as those in Pennsylvania, was most important. Concern is greatest for sulfur dioxide because it requires a complex and costly process to remove it during the burning process.

A formula of $SO_2$ emissions was devised by EPA which was expressed in terms of weight of $SO_2$ relative to the Btu present in the coal. The calculations are as follows:

Assume coal with properties of

3 percent sulfur

10,000 Btu/lb, 20 million Btu's per ton

(In order to understand the relationships of $SO_2$ to sulfur it must be remembered that the weight of sulfur is twice that of oxygen. Thus $SO_2$ formed in combustion is equal to twice the weight of sulfur contained in the coal.)

The sulfur content of the coal is:

.03 x 2000 lb = 60 lb of sulfur

Emission of $SO_2$ when coal is burned is:

2 x 60 lbs per ton = 120 lbs of $SO_2$ per ton

Which related to heat content becomes

120 lbs $SO_2$ per ton/20 million Btu/per ton = 6 lbs of $SO_2$ per million Btu.

The EPA regulations classified coal into three categories of $SO_2$ content, each subject to a different level of $SO_2$ reduction. They are:

Category 1

All coal with an $SO_2$ content of equal or less than 2 lbs/million Btu must remove 70 percent of all $SO_2$

Category 2

All coals with an $SO_2$ content of between 2 lbs/million Btu and 6 lbs/million Btu must reduce $SO_2$ content so that final emissions do not exceed 0.6 lb/million Btu. This requies a variable percent reduction of $SO_2$ by 70 to 90 percent.

Category 3

All coals with an $SO_2$ content of between 6 lbs/million Btu and 12 lbs/million Btu must reduce $SO_2$ emission by 90 percent. The maximum final emissions level, or ceiling, after 90 percent reduction is 1.2 lbs/million Btu.

These regulations in essence created two regional markets for steam coal. Most western coals and a number of coal seams primarily in eastern Kentucky would comply with the emission standards without scrubbing. Pennsylvania's coal and most other eastern and midwestern coals, can only be burned after sulfur is removed by scrubbers. Thus the establishment of emission standards has limited the development of the Appalachian and Midwestern coals. Most of Pennsylvania's coals fall in Category 2. Category 1 coals and some category 2 coals are utilized by the electric utilities with the use of scrubbers.

The emission regulations are important to coal producers for they limit the type of coals that can be mined profitably. The coal operators must have a broader knowledge of the type of coal produced for the cost of utilizing medium and high sulfur costs varies considerably. The electric power utilities are placing increasing emphasis on the cost factor in their fuel selection decisions.

## Health and Safety

For more than the past 50 years technological progress has attempted to make mining a less hazardous occupation. An impetus to these endeavors was given by the passage of the Federal Coal Mine Health and Safety Act in 1969, and amended in 1977. This act has been described as the most comprehensive industrial health and safety program ever to be legislated to promote the welfare of a single class of industrial workers. The leading causes of accidents in mines are explosions, roof and rib falls, haulage and machinery.

The implementation of the health and safety acts has been directed at the physical conditions of the coal mines. As a consequence, between 1970 and 1977 74 percent of all citations issued were in the 4 areas of electricity, ventilation, combustible materials, and fire protection. As a consequence the fatalities and injuries attributable to physical mine conditions have decreased since 1969.

In contrast, the fatalities and injuries, not attributable to mine conditions, but to human causes and work practices have increased. The mining industry feels that the implementation of the health and safety acts places the responsibility for safety on the mine operator. As a result the individual miner's responsibility for his own safety and health has been too greatly deemphasized. Miners, however, are not always willing to do what is necessary to provide for their own safety. Sometimes, they will violate safety rules, reject recommended safe procedures, and take needless chances. The reason for this include saving time, avoiding extra effort and discomfort, attracting attention, and demonstrating independence.

Because the Federal Coal Mine Health and Safety Act emphasizes enforcement of violation of physical conditions of mines the opportunity for cooperation among labor, management and the government for the total causes of fatalities and injuries have not developed. The prevention of accidents requires not only improving the physical conditions in mines but also the development of satisfactory behavioral work attitudes of labor. While safety in mining is expensive, this is one factor that must be disregarded. Advancing technology will not only reduce safety costs but also raise productivity which has been lowered in order to provide a safer mining environment.

## FUTURE

The growth of the bituminous coal industry of Pennsylvania will depend on a number of economic, political and technological considerations. It is now recognized that the industry can no longer ignore the environmental aspects of mining. However, as a response to the election of Ronald Reagan as president, the coal industry is anticipating a relaxation of regulations that have hampered coal production and kept competing fuels at artificially low prices. Possibly the greatest problem to projected increases in coal production in coming years will be the lack of an adequate skilled labor supply. The development of a well-planned program for manpower requirements is critical. Despite unresolved political and economic issues coal is the nation's most abundant fossil fuel and its place in the energy economy must be assured.

## BIBLIOGRAPHY


1. Ackerman, Bruce A. and William T. Hassler, (1980), "Beyond the New Deal: Coal and the Clean Air Act," *Yale Law Journal,* 89, 1466-1571.
2. Doyle, William S., (1976), *Strip Mining of Coal: Environmental Solutions,* Park Ridge, N.J.: Noyes Data Corporation, 352 pp.
3. Heritage, John, (June 1979), "New Coal Standards [for Coal-Burning Electric Power Plants: Issued by the United States Environmental Protection Agency]," EPA (*Environmental Protection Agency*) *Journal,* 5, 18-20.
4. Kirschten, Dick, (1979), "Politics at the Heart of the Clean Air Debate," *National Journal,* 11, 812-816.
5. ________, (1979), "The Coal Industry's Rude Awakening to the Realties of Regulation," *National Journal,* 11, 178-182.
6. Maneval, David R., (Feb.-March 1972), "Coal Mining vs. Environment: A Reconciliation in Pennsylvania," *Appalachia,* 5, 10-40.
7. McDonald L. Bruce and William H. Pomroy, (1980), *A Statistical Analysis of Mine Fire Incidents in the United States from 1950 to 1977,* Information Circular 8830, Washington, D.C.: U.S. Bureau of Mines, 42 pp.
8. Miller, E. Willard, (1970), "An Analysis of Recent Trends in the Appalachian Coal Mining Industry," *Proceedings, Economic Council, American Institute of Mining Engineers,* 221-243.
9. Miorin, A.F. and others, (1979), *Tioga River Mine Drainage Abatement Project,* Springfield, VA: Industrial Environmental Research Laboratory, 87 p.
10. Mosher, Lawrence, (1980), "Acid Rain Fallout Threatens Subsidies for Utilities that Convert to Coal: President Carter Has Proposed $10 Billion in Subsidies for Utilities that Shift from Oil to Coal, But Environmentalists Warn of More Acid Rainfall," *National Journal,* 12, 716-720.
11. Newcomb, Richard, (Jan. 1979), "Modeling Growth and Change in the American Coal Industry," *Growth and Change,* 10, 111-127.
12. ________, (May/June 1978), "The American Coal Industry," *Current History,* 74, 206-209 +.
13. Quig, Robert H., (March 13, 1980), "Coal-Derived Fuels-The Clean Fuels of the Future," *Public Utilities Fortnightly,* 105, 25-31.
14. Schmidt, Richard A. (1978), *Coal in America: An Encyclopedia of Reserves, Production and Use,* New York, N.Y.: Coal Week, McGraw-Hill Pub. Co., 458 pp.
15. Stoffel, Jennifer, (June 1980), "Acid Rain Clouds Coal's Future," *State Government News,* 23, 3-4 +.
16. Wilmoth, Roger C., (1977), *Limestone and Lime Neutralization of Ferrous Iron Acid Mine Drainage,* Springfield, VA: Industrial Environmental Research Laboratory, 95 pp.

17. United States Bureau of Mines, Division of Mining Research - Health and Safety, (1979), *Mine Health and Safety Contract Research, Development and Demonstration in Fiscal Year, 1979,* Informational Circular 8797, Washington, D.C.: 49 pp.
18. United States House Committee on Interstate and Foreign Commerce. Subcommittee on Energy and Power, (1978), *National Coal Policy Project: Hearing, April 10, 1978 on the Need to Balance the Energy Needs of the United States against the Environmental Consequences of a Renewed Emphasis on Coal Use,* 95th Congress, 2nd Session, Washington, D.C.: 347 pp.
19. United States Senate Committee on Energy and Natural Resources, (1980), *Effects of Acid Rain: Hearing: Pt. 1, May 28, 1980, on the Phenomenon of Acid Rain and Its Implications for a Natural Energy Policy,* 96th Congress, 2nd Session, Washington, D.C.: 752 pp.
20. United States Senate Committee on Environment and Public Works. Subcommittee on Environmental Pollution, (1980), *Environmental Effects of the Increased Use of Coal: Hearings March 19-April 24, 1980,* 96th Congress, 2nd Session, Washington, D.C.: 453 pp.
21. *Acid Mine Drainage in Appalachia,* (1969), Washington, D.C.: Appalachian Regional Commission, 126 pp.
22. *Annual Report on Mining, Oil and Gas and Land Reclamation and Conservation Activities,* Harrisburg, Pa.: Department of Environmental Resources. Annual reports on Pennsylvania's mining activities since 1903.
23. "Coal: Persistent Problems Delay Boom Era," (1981), *Mining Engineering,* 33, 539-561.
24. *Minerals Yearbook,* (1875 to 1976), Fuels Volumes, Washington, D.C.: Bureau of Mines, U.S. Department of the Interior. Annual volumes on bituminous coal mining in the United States.

*Chapter Thirteen*

# Impact of TMI Nuclear Accident Upon Pregnancy Outcome, Congenital Hypothyroidism and Infant Mortality

**George K. Tokuhata, Dr. P.H.,Ph.D.**
Director
Division of Epidemiological Research
Bureau of Epidemiology and
Disease Prevention
PENNSYLVANIA DEPARTMENT
OF HEALTH
Box 90, Room 1013
Harrisburg, Pa. 17108

Dr. George K. Tokuhata is Director, Division of Epidemiological Research, Pennsylvania Department of Health and adjunct Professor of Epidemiology and Biostatistics, Graduate School of Public Health, University of Pittsburgh. He is Chief of Epidemiologic Investigation of Health Impact, Governor's Fact-Finding Committee on Nuclear Power Plants, Commonwealth of Pennsylvania. He received his B.A. from Keio University, Tokyo; Ph.D. from State University of Iowa and Dr. P.H. from Johns Hopkins University. Dr. Tokuhata has published more than 60 professional articles and serves on several National health research and advisory panels and Health manpower committees.

The Three Mile Island nuclear accident of March 28, 1979 has resulted in marked social unrest locally, nationally and world-wide, particularly with respect to the health and safety aspects of nuclear energy. Subsequent to the accident the Pennsylvania Department of Health initiated a comprehensive evaluation of possible health effects of the accident upon local population. During the 10-day period of crisis, it was not possible to ascertain accurate information regarding radioactive emissions from the damaged nuclear reactor into the environment. However, the presence of rather diffuse and growing psychological disturbance in the area was apparent.

Within a short period of days following the accident, we were able to conceptualize and develop a multidisciplinary plan for a variety of different research studies specifically designed to assess the health impact of the TMI accident. Studies conceived during this critical period mostly reflected the existing epidemiological knowledge regarding biological effects of low level ionizing radiation and severe emotional stress.

I was designated by the Governor of Pennsylvania to coordinate and manage all health-related research activities relative to TMI. At the same time, a special Advisory Panel was commissioned by the Secretary of Health to oversee and guide all TMI-related studies administered by the Department of Health.

## A. PREGNANCY OUTCOME AROUND THREE MILE ISLAND

One of the most important studies developed shortly after the accident was to determine if the TMI nuclear accident has had any measurable impacts upon pregnancy outcome and infant health in the vicinity of the damaged nuclear reactor. We knew that both ionizing radiation and emotional stress can affect human reproductive process and pregnancy outcome. We also recognized that the embryo and the fetus are highly sensitive to such environmental insults, depending upon their severity, the mode of exposure, and the gestational age when exposed.

Before describing the methodology and study design in detail, let me review briefly the current state of epidemiology of pregnancy outcome, particularly in relation to radiation and stress.

### *Radiation and Pregnancy Outcome:*

Much information is now available regarding the effects of ionizing radiation on the embryo and fetus. Most of the more reliable data are derived from animal experiment; however, certain experimental findings may be applicable to the humans, at least in a qualitative sense, while recognizing inherent limitations or difficulties in such cross-species inferences.

The most significant damage from exposure to ionizing radiation results from the direct interaction of the stream of ions produced by radiation with the nucleus of the irradiated cells. The cell may be killed, the radiation may produce no damage, or such damage may be repaired. There is another type of damage which is probably the most significant one, i.e., the damaged cell survives and reproduces a clone of abnormal cells which may result in malignancies or congenital anomalies.

Possible effects of radiation on pregnancy outcome are (a) intrauterine and extrauterine growth retardation, (b) embryonic, fetal or neonatal death, and (c) gross congenital malformations. The tissue (organ) most readily and consistently affected by radiation is the *central nervous system.*

Laboratory and clinical studies by and large support the contention that doses of radiation less than 10 rads do not contribute to intrauterine or extrauterine growth retardation or to gross congenital malformations (1). Distribution of the absorbed dose from X-rays or gamma rays externally exposed is considered to be rather uniform in the developing embryo or fetus; thus, a child with multiple radiation-induced malformations is also likely to have intrauterine growth retardation and some CNS abnormalities.

To determine the effect of radiation upon pregnancy outcome, one must consider (a) the absorbed dose, (b) the dose rate (acute or chronic; continuous or intermittent), (c) the stage of gestation at which the exposure occurred, (d) the age of the mother when conceived, and (e) the health condition of the mother, in general. If the dose rate is reduced significantly, the damaged cell may recover from it in time. The pre-implanted stage of the embryo is the most sensitive to lethal effects of radiation. Embryos destroyed at this stage of pregnancy may never be recognized or recorded. However, preplantation irradiation has no apparent relationship to teratogenesis. Radiation has its greatest effectiveness in producing *congenital malformations* during the organogenesis period. In humans, this corresponds to the 14th-49th day of gestation (2).

The peak incidence of *gross malformations* occurs when the fetus is irradiated during the early organogenesis period, although cellular, tissue and organ *hypoplasia,* including *growth retardation* can be produced by radiation throughout organogenesis, and fetal and neonatal periods, if the dose is high enough. These are usually limited to CNS abnormalities and other organs, which continue to differentiate throughout gestation. Thus, *cerebral hypoplasia, microcephaly, cerebellar hypoplasia,* and *testicular atrophy* can be produced by "high" doses at specific stages of gestation.

A number of studies suggest that "low" levels of radiation with less than 10 rads of acute or chronic exposure may produce some pathologic effects in the embryo (3), (4), (5), (6), but these minor effects may be subtle and thus difficult to detect. For this reason, the National Council on Radiation Protection and Measurements has established the *maximum permissible dose* (MPO) to the fetus from occupational exposure of the expectant mother well below the known teratogenic dose. The *neonatal death* rate is highest in the surviving embryos irradiated during the early organogenesis period.

Radioactive isotopes administered internally to the pregnant woman have a variable distribution in the embryo and fetus depending upon (a) the stage of gestation, (b) whether the radioactive material crosses the placenta, and (c) the biochemical affinities of the type of radiation emitted (alpha, beta, or gamma). Thus, the evaluation of the relative risk is much more complex and difficult for radiations absorbed from internally administered radioactive materials than for radiation delivered from external X-rays or gamma ray sources.

It is generally assumed that *embryonic germ cells* are susceptible to the mutagenic effects of radiation throughout gestation. However, there is some uncertainty as to whether "low" doses below 10 R or low dose rates can produce significant cytogenetic defects. There has been no convincing evidence that cytogenetic (chromosome) abnormalities as such caused by radiation in utero have caused any significant increase in the incidence of clinical diseases.

There is no doubt that "high" doses of radiation can be carcinogenic. However, whether or not "low" doses, such as below 2 rads, can induce leukemia and/or other malignant tumors in the humans has been debated by some epidemiologists and radiation biologists. It appears improbable that radioactive fallout as reported in the past or natural background radiation as such, significantly affects the incidence of *congenital malformations, growth retardation* or *fetal death.*

The exact nature and extent of damages caused by "low" doses of radiation upon humans are still unknown. However, if the cell nucleus is damaged by radiation and some genetic materials (DNA) are lost or impaired, one may not conclude that the risk is zero. It is logical to assume then, that there is no threshold in radiation effect which may increase more or less with the increase or accumulation of exposure. However, the problem we are facing today is that anomalies caused by such "low" doses of radiation, if any, may not be detectable with the existing method of epidemiologic inquiry.

It is also important to recognize that not all persons run the same risk of developing a malignancy or other abnormalities from a given radiation exposure. These variations depend upon individual genetic-constitutional makeups, as well as different individual experiences and environmental exposures.

*Stress and Pregnancy Outcome:*

Stress or psychoemotional disturbance is considered by many researchers as a precursor to disease. There are a number of studies in humans which have found an association between prenatal anxiety/stress and gestational, perinatal and development pathology. While some of these studies seem to have methodological flaws, several have found a significant relationship to either complications of pregnancy (7), (8), or to infant growth and development (9).

Nuckolls, in particular studied the effect of "social support" upon pregnancy outcome (10). Women with a high number of "psychological assets" had one third the pregnancy and perinatal complication rate of women whose "psychological assets" were low.

Newton (11) in a retrospective study of postpartum women showed that pregnancies terminating in premature labor were more likely to have been stressful. In terms of the average number of life events per pregnancy, it was clear that the more premature the onset of labor, the higher the level of psychological stress was likely to be. The groups were matched for age, gravidity, and parity. The results of the study were independent of socioeconomic levels.

The findings of the studies cited above suggest a number of practical and scientific questions to be addressed within the context of TMI Health Effect Research Program. The *first*, and most obvious question, is whether or not the local population, including pregnant women, *as a whole* experienced any detectable stress effects. Previous studies of stress and pregnancy complications have found relationships which are either relatively weak (12) or restricted to subgroup of the overall study population.

A *second* question concerns factors which render individual women, particularly vulnerable to stress effects. As reviewed earlier, stress may be associated with morbidity only in the absence of supportive interpersonal relations. This observation is in accord with other studies of stress and illness (13), as well as with Burchfield's (14) argument that a maladaptive response to stress is atypical, and likely to occur only when adequate coping resources are unavailable.

An assessment of the role of social support, as well as other possible mediating factors, would contribute to the study of stress as a scientific concept and provide information as to which segments of the pregnant population might be at risk for stress-induced morbidity.

While specific mechanism of stress-induced morbidity is not yet fully understood, there may be several different explanations with respect to pregnancy outcome; e.g., stress-anxiety induced changes (a) in maternal behavior, such as increased smoking, drinking or medication during pregnancy, (b) in obstetric practice, such as increased prescription of analgesics and psychotropic drugs or use of special procedures, (c) in maternal-infant bonding and child-rearing practices, and (d) in the hypothalamic-adrenocortical mechanisms (15).

A carefully designed retrospective cohort study of pregnancy outcome was initiated in August, 1979 following four months of preparation. This study covered all pregnant women residing within a 10-mile radius of the TMI, who gave births during a one-year period from March 28, 1979 through March 27, 1980. This study cohort consisting of approximately 4,000 deliveries will be compared with a control cohort of another 4,000 deliveries during a one-year period immediately following the study cohort in the same geographic area. The study cohort will also be compared with similar data collected in the same general area during the immediately preceding four-year period. This design will make it possible to compare pregnancy outcome measures among three cohorts, one study group and before-and-after control groups.

Measures of adverse pregnancy outcome being investigated are: fetal deaths (stillbirths with and without abortions of 16-week or more gestation) as expressed per 1,000 deliveries, neonatal deaths (deaths within 28 days postpartum) as expressed per 1,000 live births, hebdomadal deaths (deaths within 7 days postpartum) as expressed per 1,000 live births, perinatal deaths (combined measure of fetal and neonatal deaths) as expressed per 1,000 deliveries, prematurity (gestation less than 37 weeks) as expressed in percent, immaturity (birth weight less than 2,500 grams) as expressed in percent, congenital malformations (one or more defects observed at birth) as expressed in percent, and low Apgar score (less than 7 at one minute of delivery) as expressed in percent.

As indicated earlier, the main objective of the present investigation is to determine if the TMI nuclear accident has had a measurable effect on pregnancy outcome. However, there are numerous factors other than radiation and stress that are known or suspected to influence the course of pregnancy and fetal outcome. In order to delineate the effect of the TMI accident, other known influences must be taken into account. This necessitated ascertainment of the appropriate data on a large number of variables pertaining to the pregnant women themselves and the surrounding complex environment to which they have been exposed.

The *maternal factors* considered in this study include: *sociodemographic* characteristics, such as race, age, education, occupation, employment, marital status, religion and residence; *behavioral attributes* such as smoking, drinking, and birth control practice; and *medical-obstetric histories,* such as diabetes, hypertension, thyroid disease, obesity, previous abortions-miscarriages, previous fetal deaths, prematurity/immaturity, congenital malformations, and gravidity/index birth order.

The *provider factors* that were taken into account are: *medical specialty* of the attending physician (obstetrician; general-family practitioner; osteopath; etc.); *type of practice* (solo vs. group); and *prenatal care* (initiation of medical care, frequency of visits, special procedures or tests done, instructions given, medications administered, X-ray exposures, etc.).

*Maternal stress* during the index pregnancy is being measured by overt personal statements of "anxiety-fear" as experienced by individual women during the crisis, as well as by actual stress-coping patterns, such as taking tranquilizers and sleeping pills. *Maternal radiation exposure* during the 10-day crisis following the nuclear accident is being estimated by the Department of Radiation Health of the University of Pittsburgh Graduate School of Public Health. For this purpose, all available and reliable radiation source data compiled by various agencies are being reviewed carefully, consolidated and computer analyzed on the digitized electronic maps with respect to distance and direction of each pregnant woman from the Three Mile Island. Also, added to this body of data is detailed data relative to individual whereabouts during the 10-day period so that more accurate radiation exposure can be estimated. Eventually, two series of dose estimates on an individual basis will be established, namely, *maximum possible dose* and *most likely dose.*

These *radiation dose estimates,* together with the measures of psychological stress will be related to each of the eight pregnancy outcome measures, while holding constant influences of all other factors considered in the study. The impact of the TMI nuclear accident will be assessed in terms of both radiation and stress combined, as well as each factor considered independently.

Since the level of radiation exposure is considered to be very low and thus no major radiation effect upon pregnancy outcome is expected. On the other hand, it is possible that some measurable radiation effects might be detected if the originally "reported" radiation dose data were significantly underestimated. Our ability to detect relatively small differences, if any, in the incidence of adverse consequences attributable to the nuclear accident will heavily depend upon how well we can control (take into account) the influences of all the other factors, some of which are known to be much more important than low level radiation and/or psychological stress being investigated.

## B. CONGENITAL HYPOTHYROIDISM

One of the important radioactive releases during the TMI nuclear accident was that of $I^{131}$. Since it is known that *radioactive iodine* can cause hypothyroidism and that $I^{131}$ can be taken up by pregnant women in the vicinity of TMI which, in turn, absorbed by the fetal thyroid gland through placenta after the 10th week of gestation, we decided to examine the incidence of *congenital hypothyroidism* among newborn infants. The fetal thyroid gland is much more sensitive to radioactive iodine than is the mother's thyroid gland (fetal thyroid affinity for iodine is greater than maternal thyroid affinity); i.e., a relatively small dose to the mother can be a relatively large dose to the fetus.

Beginning in July 1978, all children born alive in Pennsylvania are required to be screened for hypothyroidism, the condition characterized by lack of or insufficient level of thyroid hormone in the infant's blood. The purpose of this screening program is to find newborns with metabolic defects early enough after birth to prevent mental retardation.

The thyroid screening procedure involves testing for "low" thyroxine (T4) and "high" pituitary thyroid stimulating hormone (TSH). Confirmation of diagnosis is done through thyroid scan, which can help determine various types of the abnormality. During the initial six-month period, testing procedures were not fully standardized and the results were not considered to be complete.

There are several different diagnostic classes in congenital hypothyroidism: (a) *agenesis* (absence of the thyroid gland), (b) *dysgenesis or ectopic type* (incomplete maturation and/or displacement of the thyroid gland from the normal position), (c) *dyshormonogenesis or genetic type* (lack of enzyme necessary to synthesize thyroxine and/or difficulty in the release mechanism of thyroxine; the condition usually inherited from the parents of an autosomal reassive trait) and (d) *other types* (abnormalities caused by environmental agents). In a normal population, the incidence of congenital-neonatal hypothyroidism is in a range of one in 4,500 to 5,000 live births.

During the March 28, 1979 - March 27, 1980 period only one case of congenital hypothyroidism was identified within a ten-mile radius of TMI among approximately 4,000 newborn infants. This incidence rate is well within a normal range of expectation.

The Statewide incidence of congenital hypothyroidism for 1979 (12-month period) was one per 4,600 live births, which is also within a normal range of expectation. The rate for 1978 (only the latter 6-month period) was considerably lower; this was expected because of the fact that the thyroid screening program in Pennsylvania was started in July 1978 and that during this start-up period the screening procedures and standards were not yet fully established, making data unsuitable for comparison. The Statewide incidence for 1980 was one per 4,427 live births, again indicating that the level of congenital hypothyroidism for Pennsylvania as a whole remained within a normal range.

An apparent clustering of seven cases of congenital hypothyroidism in Lancaster County during 1979 was subjected to a special in-depth analysis and investigation because of physical proximity and timing of the Three Mile Island nuclear accident. The following diagnostic and epidemiological features are of interest: (a) One of the seven cases identified was reported in January of 1979, prior to the TMI accident, thus has no connection with radioactive iodine released from the damaged nuclear reactor. (b) One with severe multiple central nervous system anomalies was born three months after the accident; this case is unlikely to be associated with TMI accident because of the late gestation period of the fetus when the nuclear accident occurred (most, if not all, of these defects would have come about prior to the TMI accident) and also of coexisting developmental anomalies which are unlikely to be associated with radiation. (c) One case was of dysgenesis, representing one of discordant Amish twins, thus, non-supportive of the etiology secondary to radiation exposure. (d) Another case of dysgenesis in whom the thyroid glands were displaced from the normal position. (e) One case of dyshormonogenesis from an Amish family where the condition (lack of enzyme to synthesize thyroxine) was inherited from the parents. (f) For the remaining two cases no thyroid scan was conducted.

Having completed detailed diagnostic analysis and epidemiological assessment of all the cases reported in Lancaster during 1979, we concluded that cases of congenital hypothyroidism were not related to the TMI nuclear accident. Except for the two cases for which diagnostic scan was not performed (unknown type), these types of anomalies are not expected to result from direct or indirect exposure of the fetus to radioiodine. This conclusion was also supported by an independent Hypothyroidism Investigative Committee organized by the State Health Department, which included expertise in the fields of epidemiology, pediatric endocrinology, obstetrics, medical genetics, biostatistics, and radiation physics.

Apart from the incidence analysis presented above, there is also an important consideration with respect to radiation in relation to congenital hypothyroidism.

*First,* after March 28 through December 31, 1979, no single case of congenital hypothyroidism was reported in Dauphin, Cumberland, Perry, Northumberland, Juniata, Snyder, Mifflin, and Union Counties, the areas downwind (N, NW, NNW) from the Three Mile Island during the first 48 hours of the accident, when probably the largest amount of radioactive releases took place, thus the largest amount of contamination including $I^{131}$.

*Second,* the maximum combined (inhalation and ingestion) *human thyroid dose* of radioactive iodine in the vicinity of the TMI following the March 28, 1979 accident through April 1979 is estimated to be 7.5 mrad (Editorial: Annals of Internal Medicine, Vol. 91, No. 3, September 1979). At least 1,000 times greater thyroid doses (i.e., 7.5 rads) would be required to have significant acute damages to the thyroid gland; however, even at this dose level, many of the damaged cells may be repaired. Based on the experiences of the Marshallese exposed to fresh radioactive fallout and atomic bomb victims, it is considered likely that as much as 50 to 100 rads fetal thyroid doses would be necessary to cause irreversible tissue damages, such as congenital hypothyroidism and/or thyroid cancer. Acknowledging the fact that the *fetal thyroid* is much more sensitive to radioiodine than is the *maternal thyroid* (a conservative upper bound estimate is that the thyroid dose to a fetus may be as high as ten times the maternal thyroid dose), the maximum likely *fetal thyroid dose* (approximately 75 mrad) and the maximum possible thyroid dose of 190 to 200 mrad in the vicinity of the damaged nuclear plant are still far too small to have caused congenital hypothyroidism.

In an epidemiological investigation of possible "clustering" of a disease or morbid condition, it is important to recognize the technical difficulty and methodological limitations associated with such investigation. It is the overall consistent pattern of observation that provides useful basis for conclusion, rather than a single isolated change or difference, which in most cases occurs without substantive epidemiologic significance. This is particularly true when relatively small populations are being studied. One may or may not find a "statistically significant" change, difference, or clustering in morbid rates in an area, depending upon how such population is delineated geographically and/or temporarily. It is equally important that investigators carefully examine the observed relationships and determine if such relationships are consistent with the known biological theory or orientation, which is based on the previous studies and experiences. Our conclusions regarding congenital hypothyroidism around the Three Mile Island nuclear plant have been based on both the overall pattern of epidemiologic observations and in reference to the existing scientific knowledge.

## C. INFANT MORTALITY

Ionizing radiation is often related to infant morbidity and mortality in a general context of biological effects because of the greater sensitivity of the newborns to radiation, as compared with the adult population. The infant mortality is defined

TABLE 1
*Resident Live Births by Quarter:*
*Pennsylvania and Ten Mile TMI Area Communities, 1977-1979*

| Year/Quarter | Pennsylvania | TMI Ten Mile Area Total | Harrisburg City | Excluding Hbg. City |
|---|---|---|---|---|
| 1977 | (153,415) | (3,750) | (1,001) | (2,749) |
| Jan.-March | 36,911 | 886 | 242 | 644 |
| April-June | 38,414 | 937 | 248 | 689 |
| July-Sept. | 40,181 | 977 | 274 | 703 |
| Oct-Dec. | 37,909 | 950 | 237 | 713 |
| 1978 | (151,438) | (3,803) | (1,057) | (2,746) |
| Jan.-March | 37,084 | 926 | 261 | 665 |
| April-June | 36,339 | 922 | 262 | 660 |
| July-Sept. | 39,932 | 1,029 | 302 | 727 |
| Oct.-Dec. | 38,083 | 926 | 232 | 694 |
| 1979 | (157,533) | (3,905) | (1,185) | (2,720) |
| Jan.-March | 38,326 | 932 | 296 | 636 |
| April-June | 38,351 | 983 | 303 | 680 |
| July-Sept. | 41,933 | 1,023 | 302 | 721 |
| Oct.-Dec. | 38,923 | 967 | 284 | 683 |

as the risk of infants dying within the first year of life and is expressed per 1,000 live births.

Subsequent to the March 1979 nuclear accident, we initiated a comprehensive evaluation of the existing vital statistics data in order to determine if the TMI accident has had any measurable influence upon infant mortality in the vicinity of the damaged plant.

For the purpose of the present study, we considered a 10-mile radius of the Three Mile Island, wherein approximately 4,000 infants are born annually (Table 1). Both levels of radiation exposure and psychological distress within the 10-mile radius communities were higher than those beyond the 10-mile radius communities. The available mortality data were analyzed by calendar quarters, as well as annually, for each of the three consecutive years, 1977, 1978, and 1979, for the entire 10-mile area, including Harrisburg, the 10-mile area excluding Harrisburg, and Harrisburg separately. For cross-sectional comparison, corresponding mortality data for the State of Pennsylvania as a whole, were evaluated for the same historical time frames. As indicated in Table 2, the *infant mortality rate* was not significantly different between the 10-mile area with or without Harrisburg and the State of Pennsylvania for any of the three years under consideration. The higher infant death rate indicated for Harrisburg separately is a reflection of the fact that approximately one-half of the infants born in the city were nonwhite.

The infant mortality rate within the 10-mile radius, including Harrisburg, was already considerably high (19.3 per 1,000 live births) during the *first quarter* of 1979 prior to the TMI accident. The rate remained at the same level during the *sec-*

TABLE 2

*Resident Infant Deaths, Number and Rate, by Quarter:*
*Pennsylvania and Ten Mile TMI Area Communities, 1977-1979*

| Year/Quarter | Number of Deaths | | | | Death Rate Per 1,000 Live Births | | | |
|---|---|---|---|---|---|---|---|---|
| | | Ten Mile TMI Area | | | | Ten Mile TMI Area | | |
| | Pa. | Total | Harrisburg City | Excluding Hbg. City | Pa. | Total | Harrisburg City | Excluding Hbg. City |
| | | | | Infant Deaths | | | | |
| 1977 | (2,137) | (47) | (15) | (32) | (13.9) | (12.5) | (15.0) | (11.6) |
| Jan.-March | 544 | 13 | 6 | 7 | 14.7 | 14.7 | 24.8 | 10.9 |
| April-June | 554 | 11 | 2 | 9 | 14.4 | 11.7 | 8.1 | 13.1 |
| July-Sept. | 520 | 9 | 3 | 6 | 12.9 | 9.2 | 10.9 | 8.5 |
| Oct.-Dec. | 519 | 14 | 4 | 10 | 13.7 | 14.7 | 16.9 | 14.0 |
| 1978 | (2,031) | (41) | (18) | (23) | (13.4) | (10.8) | (17.0) | ( 8.4) |
| Jan.-March | 530 | 13 | 8 | 5 | 14.3 | 14.0 | 30.7 | 7.5 |
| April-June | 509 | 9 | 3 | 6 | 14.0 | 9.8 | 11.5 | 9.1 |
| July-Sept. | 473 | 5 | 1 | 4 | 11.8 | 4.9 | 3.3 | 5.5 |
| Oct.-Dec. | 519 | 14 | 6 | 8 | 13.6 | 15.1 | 25.9 | 11.5 |
| 1979 | (2,118)* | (63) | (31) | (32) | (13.4) | (16.1) | (26.2) | (11.8) |
| Jan.-March | 511 | 18 | 10 | 8 | 13.3 | 19.3 | 33.8 | 12.6 |
| April-June | 537 | 19 | 9 | 10 | 14.0 | 19.3 | 29.7 | 14.7 |
| July-Sept. | 507 | 13 | 3 | 10 | 12.1 | 12.7 | 9.9 | 13.9 |
| Oct.-Dec. | 562 | 13 | 9 | 4 | 14.4 | 13.4 | 31.7 | 5.9 |

*Includes one death, month of occurrence, unknown.

TABLE 3

*Resident Fetal Deaths (Total), Number and Rate, by Quarter: Pennsylvania and Ten Mile TMI Area Communities, 1977-1979*

| Year/Quarter | Number of Deaths | | | | Death Rate Per 1,000 Deliveries* | | | |
|---|---|---|---|---|---|---|---|---|
| | | Ten Mile TMI Area | | | | Ten Mile TMI Area | | |
| | Pa. | Total | Harrisburg City | Excluding Hbg. City | Pa. | Total | Harrisburg City | Excluding Hbg. City |
| | | | | Fetal Deaths (Total) | | | | |
| 1977 | (4,058) | (80) | (45) | (35) | (25.8) | (20.9) | (43.0) | (12.6) |
| Jan.-March | 1,062 | 16 | 10 | 6 | 28.0 | 17.7 | 39.7 | 9.2 |
| April-June | 992 | 18 | 9 | 9 | 25.2 | 18.8 | 35.0 | 12.9 |
| July-Sept. | 1,026 | 23 | 12 | 11 | 24.9 | 23.0 | 49.0 | 15.4 |
| Oct.-Dec. | 978 | 23 | 14 | 9 | 25.1 | 23.6 | 55.8 | 12.5 |
| 1978 | (4,034) | (77) | (38) | (39) | (25.9) | (19.8) | (34.7) | (14.0) |
| Jan.-March | 1,003 | 15 | 5 | 10 | 26.3 | 15.9 | 18.8 | 14.8 |
| April-June | 1,047 | 21 | 12 | 9 | 28.0 | 22.3 | 43.8 | 13.5 |
| July-Sept. | 1,001 | 20 | 10 | 10 | 24.5 | 19.1 | 32.1 | 13.6 |
| Oct.-Dec. | 983 | 21 | 11 | 10 | 25.2 | 22.2 | 45.3 | 14.2 |
| 1979 | (3,608) | (67) | (37) | (30) | (22.4) | (16.9) | (30.3) | (10.9) |
| Jan.-March | 938 | 24 | 13 | 11 | 23.9 | 25.1 | 42.1 | 17.0 |
| April-June | 916 | 12 | 6 | 6 | 23.3 | 12.1 | 19.4 | 8.7 |
| July-Sept. | 937 | 16 | 10 | 6 | 21.9 | 15.4 | 32.1 | 8.3 |
| Oct.-Dec. | 817 | 15 | 8 | 7 | 20.6 | 15.3 | 27.4 | 10.1 |

*Deliveries: Live births and fetal deaths (including abortions).

*ond quarter* of 1979 immediately following the accident, but declined substantially during the *third* (12.7) and *fourth* (13.4) *quarters.* This temporal pattern of change in the rate is consistent with the view that the TMI accident has had no measurable impact upon infant mortality. Otherwise, the infant mortality rate would have increased steadily (or, at least, would have remained high as a result of interaction between seasonal downward trend and TMI-related upward trend), particularly during the third and early fourth quarters. Fetal sensitivity to radiation and maternal distress is much greater in the earlier period of gestation or organogenesis when exposed and this would have been reflected on the gradually rising mortality trend following the accident for a period of nine to ten months. However, the actual observation was contrary to this hypothesis.

Within the 10-mile radius of TMI, the 1979 infant mortality rate (16.1) was not significantly different from the 1977 rate (12.5). The 1978 infant mortality rate (10.8) in the same area was somewhat atypical and unusually low, particularly within the immediately surrounding communities outside of Harrisburg (8.4). This is largely because of the small population, wherein marked statistical variations from year to year are not at all uncommon with no particular epidemiologic significance. For this reason, the 1978 infant mortality rate should not be used as a normal base for comparison.

Having considered both cross-sectional and temporal analyses of the available vital statistics data compiled by the State Health Department, we found no evidence that the TMI nuclear accident has had any significant impact upon infant mortality. Statistical variations or differences, as observed in the 10-mile radius, are considered to be a typical random phenomenon in a relatively small population with no particular epidemiologic significance. Theoretically, too, the low levels of radiation exposure, as reported offsite, cannot be directly related to such massive destruction or impairment of cells that cause infant deaths.

The pattern of *fetal mortality rate* or the risk of the fetus being born dead in the vicinity of the Three Mile Island was also analyzed by the same method as applied to infant mortality. We found that there is no indication that the TMI nuclear accident was related to its quarterly or annual variations in fetal mortality (Table 3). The level of fetal mortality within the 10-mile communities was, in fact, considerably lower than that for the State as a whole.

## SUMMARY AND CONCLUSIONS

The Three Mile Island nuclear accident has caused an extensive social and political unrest world-wide. At the same time, it has presented social scientists and biomedical investigators a unique opportunity to evaluate its impact upon local population. Probably the most important concern is that of safety and health effects of this unprecedented event.

From the currently available epidemiological knowledge, no significant physical health effects are expected from the low level radiation reported to have been released from the damaged TMI facility. However, in the absence of absolute certainty as to the exact amount of radioactive contamination of the local environment and population, particularly during the early period following the accident, carefully designed epidemiological studies, such as those described in this report, are justified. Furthermore, some substantial psychological impacts upon local populace have been documented. It is not known at this time what significant physical manifestations, if any, may actually ensue from psychological distress over an extended period of years. In addition, because of high sensitivity of the fetus to ionizing radiation and severe maternal stress, timely evaluation of pregnancy outcome should be pursued.

Based on the already established TMI Population Registry, there should be a continuous and long-term epidemiologic surveillance of the exposed general population, which includes approximately 37,000 individuals. The TMI Population Registry is now updated annually so that annual mortality (rate and cause) and periodic health survey, can be conducted. Such endeavor will make it possible to determine and document if there is any measurable health impact in humans from the low level of ionizing radiation that has not yet been fully studied.

It is the responsibility of government agencies and academic communities to properly inform the general public with the objective results of carefully-designed scientific studies of possible health effects of the TMI nuclear accident. It is also critical that safety of nuclear energy is properly addressed in relation to that of other available means of energy production, as well as many other potential risks in human life. Equally important is the public understanding of various alternatives, so that the society as a whole, rather than selected few, can make a rational choice for its constituents. It is hoped that TMI Health Effect Research Program will serve as a means to achieve such goals.

## REFERENCES


1. Brent, R.L. and Gorson, R.O.: Radiation Exposure in Pregnancy. Current Problems in Radiology, Vol. II, No. 5, 1972.
2. Brent, R.L.: Effects of Ionizing Radiation on Growth and Development. Contributions to Epidemiology and Biostatistics, Vol. 1, 1979.
3. Meyer, M., Diamond, E. and Merz, T.: Sex Ratio of Children Born to Mothers who had been Exposed to X-rays in Utero. Johns Hopkins M.J. 123: 123, 1968.
4. Segall, A., MacMahon, B., and Hannigan, M.: Congenital Malformations and Background Radiation in Northern New England. J. Chron. Dis. 17:915, 1964.

5. Tabuchi, A.: Fetal Disorders Due to Ionizing Radiation, Hiroshima J. M. Sc. 13:125, 1964.
6. Kinlen, L.J. and Acheson, E.D.: Diagnostic Irradiation, Congenital Malformations and Spontaneous Abortions, Brit. J. Radiol. 41:648, 1968.
7. Nuckolls, K.B.: Psychological Assets, Life Crisis and the Prognosis of Pregnancy. American J. Epid. 95:431, 1972.
8. Morishima, H.O.: The Influence of Maternal Psychological Stress on the Fetus, Amer. J. Obs. and Gyn. 131:286, 1978.
9. Barlow, S.M.: Delay of Postnatal Growth and Development of Offspring Produced by Maternal Restraint Stress During Pregnancy in the Rat. Teratology 18:211, 1978.
10. Nuckholls, K.B.: op. cit.
11. Newton, R.W.: Psychosocial Stress in Pregnancy and its Relation to the Onset of Premature Labor. Brit.Med.J. 2:411, 1979.
12. Gorsuch, R.L. and Kay, M.K.: Abnormalities in Pregnancy as a Function of Anxiety and Life Stress. Psychosomatic Med. 36:352, 1974.
13. Berkman, L.F. and Syme, S.L.: Social Networks, Host Resistance, and Mortality. American J. Epid. 109:186, 1979.
14. Burchfield, S.R.: The Stress Response: A New Perspective. Psychosomatic Med. 41:661, 1979.
15. Smith, D.J.: Modification of Prenatal Stress Effects in Rats by Adrenalectomy, Dexamethasone, and Chlorpromazine. Physiology and Behavior 15:461, 1975.
16. Houts, P.S., Miller, R.W., Tokuhata, G.K., and Ham, K.S.: Health-Related Behavioral Impact of the Three Mile Island Nuclear Incidence. Report submitted to the TMI Advisory Panel on Health Research Studies, Pennsylvania Department of Health. Part I. April, 1980.

*Chapter Fourteen*

# Psychological and Social Effects on the Population Surrounding Three Mile Island After the Nuclear Accident on March 28, 1979


| Peter S. Houts, Ph.D. | Marilyn K. Goldhaber, M.P.H. |
| --- | --- |
| The Milton S. Hershey Medical Center PENNSYLVANIA STATE UNIVERSITY College of Medicine Hershey, Pa. 17033 | PENNSYLVANIA DEPARTMENT OF HEALTH |


Dr. Houts received a B.A. degree from Antioch College and a Ph.D. in Social Psychology from the University of Michigan. He has taught at Goucher College and was a post-doctoral research fellow in the Psychiatry Department of the Stanford University Medical School before joining the faculty of the Department of Behavioral Science of The Pennsylvania State University College of Medicine in 1967. He is now associate professor in that department. He has directed studies on the Health Related Behavioral Impact of the Three Mile Island Nuclear Incident under contracts with the Pennsylvania Department of Health.

Ms. Goldhaber graduated from the University of California, Berkeley, with a B.A. in Mathematics and an M.P.H. in Biostatistics. She has worked in cancer research at the Resource for Cancer Epidemiology in Berkeley, California and is currently employed by the Division of Epidemiological Research at the Pennsylvania Department of Health. The Division of Epidemiological Research is conducting a comprehensive Three Mile Island Research Program where Ms. Goldhaber is in charge of the Three Mile Island Population Registry.

The crisis at Three Mile Island (TMI) nuclear power plant began on March 28, 1979. It lasted a little more than a week and caused the temporary exodus of a substantial portion of the population in the vicinity of the plant. During the crisis many area residents were fearful of their safety and many who evacuated were not sure that they would ever be able to return to their homes. After the immediate crisis passed, there was a continuing controversy about how to deal with the damaged reactor. As a result, the situation at Three Mile Island received almost continuous attention in national and local news media for over a year after the original accident. What began as a short term crisis became, for many persons in the area, a chronically disturbing situation.

This chapter is concerned with social, psychological and economic impacts of the TMI crisis on persons and institutions in the vicinity of the reactor. Studies dealing with the decision to evacuate, the economic impact of the crisis, the effect of the crisis on mobility of persons living near the plant, attitudes toward TMI following the crisis and psychological stress experienced by people in the vicinity of the facility will be reviewed. Studies in each of these areas were carried out either directly by the Pennsylvania Department of Health or under its auspices and these studies will be the primary focus of this chapter. Other studies carried out under auspices of the Presidential Commission on Three Mile Island, the Nuclear Regulatory Commission and the National Institute of Mental Health will also be reviewed.

*Evacuation*

Voluntary evacuation in the vicinity of Three Mile Island began almost immediately after the discovery of the nuclear accident was broadcast to the public. During the first two days after the accident the evacuation was minimal with only 5 or 6 percent of the population within the 5 mile radius leaving for ostensibly safer ground. On the third day, however, nearly 50 percent of the population within a 5 mile radius had evacuated (7). This was largely precipitated by the Governor of Pennsylvania's advisory on Friday, March 30, 1979, for voluntary evacuation of pregnant women and small children residing within 5 miles of the facility. This, compounded by the increasingly confusing and sometimes alarming stories being reported by news media and the fact that the weekend was approaching, led to the exodus of an estimated 39% of the total residential population within 15 miles of the plant by the weekend. (This involved 60% of the population in a 0-5 mile ring, 44% of the population in the 5-10 mile ring and 32% of the population in a 10-15 mile ring.) The total number of evacuees was estimated to be almost 150,000 persons. The median distance traveled by evacuees was 100 miles and the average evacuation period was 5 days (3).

At no time did the evacuation take on the tenor of mass panic. That is, most families left in a relatively orderly fashion in their family cars at a general flow which did not noticeably disrupt normal traffic patterns. Since there was no official order other than the Governor's "advisory" the choice to evacuate was an individual or family decision. Only one community, Middletown, prepared and distributed hand bills on the fifth day after the accident depicting plans for a possible evacuation. A Red Cross sponsored shelter for women and small children was set up in the town of Hershey (approximately 10 miles from the plant) to accommodate persons from the five mile radius who came under the Governor's advisory but had no family or friends with which to stay. It was later found that 85% of all evacuees stayed with family or friends and only 15% were obliged to find hotel or motel accommodations (3).

Evacuation behavior, assessed through surveys of the population after the crisis, showed that the older the head of the household, the less likely that the household evacuated. Families with small children, higher than average educational attainment, and higher than average income were more likely to evacuate. The presence of a pregnant woman also had a significant affect on the decision to leave the area (3).

In general women were more prone to evacuate than men. Younger persons were more likely to evacuate than older persons. Young parents (approximately in their 20's) and their small children were more likely to stay away for longer periods than were other age groups. Evacuation patterns also differed slightly by geographic location. This was especially apparent in several small communities where the perception of distance to the plant differed substantially from actual distance.

A survey in July 1979 (7) asked respondents who had left the area during the crisis their reasons for evacuating. The most common reasons were that the situation seemed dangerous (82%) and that the information available was confusing (78%). Those who had stayed during the crisis were also asked their reasons for staying. The most frequently given reasons were: "whatever happens is in God's hands" (70%) and "waiting for evacuation order" (62%). Interestingly only 30% of the stayers said they saw no danger.

## ECONOMIC IMPACT

The accident at TMI had an immediate, detrimental impact on the economy of the area surrounding the plant according to a report of the Governor's Office on Policy Planning (6). However the impact was short-lived and the economy of South Central Pennsylvania returned to nearly normal levels within 2 weeks after the accident. Some specific aspects of the economy, notably food processing (dairy), real estate and tourism took longer to return to normal. However, within 3 months the economy, other than the nuclear industry, seemed imperceptably changed if at all. (Effects of the crisis on the nuclear industry will not be addressed here.)

The costs of evacuation was the greatest source of economic loss for both individuals and communities. With 39% of the population residing within 15 miles of the plant temporarily away, the routine of economic life was drastically altered. Retail sales were reduced, industrial production was interrupted and normal business patterns were disrupted. Schools were closed, conventions had to be cancelled, and, in some of the communities near the plant, curfews were enforced.

Short term losses to business were estimated, in the report of the Governor's Task Force on Three Mile Island, at 7.7 million in the value of production and manufacturing, 74.2 million in business sales and non-manufacturing and .25 to .50 million in agriculture (13). However, both tourism and housing were able to rebound to normal levels and even make up for their temporary losses (adjusting for fuel price increases, high interest rates and shortage of mortgage funds). Thus, the total economic loss to businesses in the area was estimated to be about 82 million dollars.

Two other studies assessed personal costs (direct costs to citizens) in terms of evacuation expenses, wages lost, additional drug-alcohol, cigarette consumption expenditures (due to stress) and increased health facility usage. These studies were sponsored by the Nuclear Regulatory Commission and the Pennsylvania Department of Health (3, 11). They found that costs of evacuation were substantial to many households, averaging about 200 dollars per evacuated household. Subtracting out insurance reimbursements, the total net cost to the residents within a 15 mile radius around the plant was estimated at 6-10 million dollars. These estimates include actual evacuation expenses, pay loss and other incidental costs to residential families.

Hu et al (11) carried the investigation further by exploring possible health-related economic costs resulting from changes in physical or mental health status and/or a change in the use of health care services. They analyzed the situation from both the individual perspective (population surveys) and provider perspective (physician survey and health insurance reimbursement data). There was some indication from the population survey that increased stress may have brought about a small increase in health care utilization (physician visits). However, surveys of local physicians and studies of Blue Cross/Blue Shield and Medicare/Medicaid data indicated little, if any, increase in health care utilization attributable to the crisis, at Three Mile Island. Houts et al (10) in a study of utilization patterns among patients in a primary care family practice near Three Mile Island, came to similar conclusions. They found no increase in practice utilization that could be attributed to the Three Mile Island crisis. Interestingly, though, they did find that people who were high practice utilizers prior to the crisis tended to be more upset during the crisis than were low utilizers. However, this concern apparently did not cause the high utilizers to increase utilization beyond their baseline rates following the crisis.

## MOBILITY

A telephone survey of residents within 5 miles of TMI in July, 1979, three months after the crisis, had found that 17% of respondents living within 5 miles of TMI said that someone in their household had "considered" moving because of the situation at Three Mile Island and 6% reported that someone had definitely "decided" to move because of that situation. It therefore seemed likely that a substantial number of people might subsequently move away from the area because of the Three Mile Island crisis. However, interview data of this sort are not definitive since they only deal with intentions to move rather than actual behavior. Furthermore, it is not known whether these people might have moved in any case with or without the Three Mile Island crisis. Because of this, a study was undertaken by the Pennsylvania Department of Health in collaboration with Pennsylvania State University to examine mobility rates for the area within 5 miles of Three Mile Island during the one year period immediately following the Three Mile Island crisis in March, 1979 (5). Three questions were addressed in this study:

1. Was the percent of persons within 5 miles of Three Mile Island who moved during the year following the crisis different from percent of movers in other comparable groups during this time period?
2. For persons within 5 miles of TMI at the time of the accident, how great a role did the situation at the Three Mile Island play in the decision to move or stay, or the choice of how far to move?
3. Were the people who moved out of the area within 5 miles of TMI in the year following the crisis different from those who moved into the area?

It was possible to answer these questions by using the Three Mile Island population registry, a 95% complete registry established by the Pennsylvania Department of Health of all persons residing within 5 miles of the nuclear plant shortly after the March 28, 1979 accident. The registry was developed as part of a long-term study of possible health effects of the nuclear accident and thus required yearly updating the current addresses of all persons included therein. As a bi-product of the yearly update, residential mobility could be systematically monitored. In addition, samples of movers were identified after the first year and interviews were conducted with random samples of in-migrants, out-migrants, and non-migrants and a control group from communities 40 to 55 miles away.

The first question studied was whether the mobility rate in the vicinity of TMI was different from that of other comparable groups. The post-TMI annual mobility rate (percent of persons who moved to another dwelling) for the 5 mile community was computed to be 11.6%. This was lower than the 13.6% mobility rate for the same population during the one year prior to the Three Mile Island crisis and was the reverse of conjectures that the TMI crisis precipitated mass migration out of the area. This drop of almost 2% was, however, difficult to interpret because of a sharp escalation of interest rates and a shortage of mortgage loan funds during the year studied. Therefore, a second comparison was made with the mobility rate for the control group living from 41 to 55 miles from Three Mile Island. The rate for that group was 12.5%, only 1% greater than for the TMI area. The difference between this rate and the TMI mobility rate was evaluated statistically using analysis of covariance controlling for variables which are frequently associated with mobility: age, education, how many times the respondent had moved in his/her lifetime, whether the respondent owned or rented and marital status. The difference between the adjusted means was less than 1% and was far from statistical significance. The conclusion drawn from these analyses was that there was no conclusive evidence that mobility rates within the 5 mile radius of Three Mile Island were significantly effected by the crisis in March of 1979.

The second question addressed was how great a role the situation at Three Mile Island played in the decision to move or in how far to move. This question was addressed utilizing multiple regression analysis. The conclusion from these analyses was that, after traditional predictors of mobility (age, education, own-rent, etc.) were controlled, people's feelings about Three Mile Island had only a small effect of mobility. However, there was some evidence that, for very mobile people (young and educated) attitudes toward Three Mile Island did relate to the distance moved. The more concerned these people were about the situation at TMI, the further they moved from the facility.

The third question addressed in the mobility study was whether the type of person moving into the area differed from those moving out and, therefore, whether population characteristics of the area were changing. Once again, results were largely negative. Only two differences in demographic characteristics were statistically significant between the two groups. The age of the oldest household member for in-migrants was slightly younger than for out-migrants and there was a significantly larger percentage of TMI employees among in-migrants compared to out-migrants. There were, however, consistent differences between in- and out-migrants in their attitudes toward the situation at Three Mile Island. Five attitude measures dealing with perceived safety, dangerousness of krypton venting, being upset about Three Mile Island, support for re-starting TMI Unit One, and whether there should be more or fewer nuclear plants in the future all showed more positive TMI attitudes among in-migrants than among out-migrants.

The general conclusion from this study on mobility was that the TMI crisis had very small, if any, effect on mobility in the immediate area of the facility. This conclusion is essentially the same as that of three other studies, one conducted by the Pennsylvania Governor's Office (6), one conducted by a local board of realtors (14) and another under sponsorship of the Nuclear Regulatory Commission (4). All of these studies showed a marked drop in real estate sales in the vicinity of Three Mile Island immediately following the crisis but a return to normal levels within a month, and for the year as a whole, very little change in real estate transactions.

## ATTITUDES TOWARD THREE MILE ISLAND

Following is a summary of attitudes from surveys carried out by the Pennsylvania State University and the Pennsylvania Department of Health in July, 1979, January, 1980, and October, 1980 (7, 8, 9). All three surveys included persons living within 5 miles of TMI as well as a control group from 41-55 miles away. Each survey addressed slightly different issues related to attitudes toward Three Mile Island. Some questions were repeated in more than one survey making it possible to examine changes over time. Responses to these questions will be discussed first.

The first question concerned restarting Reactor #1, the undamaged reactor, at Three Mile Island. In January, 1980, when the question was first asked, 60% of respondents within 5 miles of Three Mile Island and 30% of respondents 41-55 miles away opposed restarting the facility. However, in October, 1980, nine months later, opposition to re-opening had decreased in the area close to the plant with only 46% of respondents within 5 miles opposing restarting compared to 42% of respondents living from 41-55 miles away opposing the restart. The decrease from 60% to 46% among the close group was statistically significant, but the increase from 30% to 42% among the control group was not. The difference between the close and far groups was statistically significant in January, 1980, but was not so in October, 1980.

The second question repeated from earlier surveys concerned political activity and asked respondents whether they had, personally, been active in any organization or had gone to any meeting to influence what happened at Three Mile Island. Results showed that in January, 1980, 13% of respondents gave affirmative answers to this question and in October, 1980, 15% did so indicating little change over the 9 month interval. It should be noted that this participation rate near Three Mile Island was high by the usual standards of political activity in this country. For example, a survey by the National Opinion Research Center in 1973 found that only 5% of their respondents had ever participated in any kind of anti-war or pro-war demonstration and that only 9.5% had ever been involved in picketing in the course of a labor dispute. The figures for Three Mile Island are even more impressive when one considers that the National Opinion Research Center questions referred to activities over an entire lifetime, where political activity related to the TMI accident could have occurred only within a period of less than 18 months.

In October, 1980, respondents were asked how much influence they felt different groups should have in the decision to clean up Three Mile Island. Strongest support was given to the Nuclear Regulatory Commission and energy experts working for the Pennsylvania Department of Environmental Resources. It appears, from this, that the public prefers technical experts working under government auspices to have the most influence in cleaning up the facility.

Respondents, in October, 1980, were also asked how strongly they believed or disbelieved a number of common rumors concerning the effects of the Three Mile Island accident. Five rumors were read to respondents including alleged increased numbers of miscarriages, increased birth defects, expectation of higher cancer rates, increased health problems in farm animals, increase in general health problems, and increase in mental health problems because of the Three Mile Island crisis. Results showed approximately equal numbers of believers and nonbelievers. The most widely accepted rumor was that there had been an increase in mental health problems because of Three Mile Island (58%), the least accepted was that there had been an increase in birth defects since the crisis (31%). For two rumors, increase in miscarriages and increases in birth defects, there was greater acceptance of the rumor among the group far from Three Mile Island than among the group within 5 miles of the facility.

Attitudes toward media coverage of the Three Mile Island situation were also assessed in October, 1980. Results showed that almost half the respondents felt that the media had blown events out of proportion and that 1/5 thought that information was withheld or covered up. Less than 1/3 felt that the events were reported accurately.

Finally, three questions were included in the October survey which dealt with attitudes toward the venting of krypton gas at the Three Mile Island facility in July, 1980. Although the authorities proclaimed it as safe, the venting of this gas had been the subject of considerable public debate in the months prior to its occurrence. Partly as a result of this debate, the public was well informed in advance of the venting. This gave persons living in the area an opportunity to evacuate during venting periods if they so wished. Fifteen percent of the population living within 5 miles of Three Mile Island in July, 1980, reported that venting was "an important reason" for their leaving the area during the venting period. The average length of absence due to the venting was 10 days. Respondents were also asked if they felt the venting was the right way to get rid of the krypton gas and also how dangerous they thought the krypton venting was. These questions were asked of respondents within 5 miles of Three Mile Island as well as the control group from 41-55 miles away. The results showed that people living close to Three Mile Island were significantly less negative about the venting than were people further away. This is consistent with the finding mentioned above where some rumors tended to be accepted to a greater degree further from the plant than among persons close to the facility.

These attitude studies indicate that a substantial percent of the population near Three Mile Island had serious concerns about the nuclear facility as long as 18 months after the crisis. However, there was also evidence that these concerns are decreasing. This is consistent with the studies of stress discussed in the next section.

## STRESS DURING AND FOLLOWING THE TMI CRISIS

One of the most widely studied aspects of the Three Mile Island crisis was the psychological stress experienced by people in the vicinity of TMI. Results of these studies, which indicate some degree of continuing distress for up to a year following the crisis, have attracted considerable public attention. In this section we will discuss primarily results of studies carried out by the Pennsylvania State University in collaboration with the Pennsylvania Department of Health (7, 8, 9). Relationships between the findings of the Penn State-Pennsylvania Department of Health studies and those of other investigators will also be discussed.

Houts et al in a 1981 report summarized the results of three surveys conducted over an 18 month period (9). These surveys were of representative samples of households with phones within 5 miles of Three Mile Island as well as in communities from 41 to 55 miles from the facility. Five distress indices were included as follows:

1. How upset the respondent was about the situation at Three Mile Island.
2. How serious a threat the respondent felt TMI was to safety.
3. Frequency of "behavioral symptoms" during a two week period (i.e., lack of appetite, overeating, sleeplessness, feeling shaky, trouble thinking, irritability and anger).
4. Frequency of "somatic symptoms" (i.e., stomachaches, headaches, diarrhea, frequent urination, rash, abdominal pain, and sweating spells) during a 2 week period.
5. For those persons who reported either behavioral or somatic symptoms, whether they attributed those symptoms to the situation at Three Mile Island.

Results, shown in Table 1, are as follows.

1. Upset ratings were significantly higher close to TMI compared to farther away from April, 1979, through January, 1980, but were no longer significantly different in October, 1980. In examination of the mean ratings at each time period shows that levels of upset came down over time for both the close and far groups, but that the drop was sharper for the groups close to TMI.

2. Both behavioral and somatic symptoms were reported more frequently close to TMI compared to 41 to 55 miles away in April and July, 1979, as well as January, 1980. However, by October, 1980 this difference was sharply reduced and the difference between the two groups was no longer statistically significant. It should be noted in interpreting these findings that the general levels of symptom reporting fluctuated considerably over this time period. This could have been due to many factors including seasonal variations. Therefore, differences between close and far groups are better indicators of the effects of Three Mile Island than are general levels of symptom reporting at each time period.

3. Perceptions of TMI as a serious threat to one's family's safety came down

TABLE 1

*Mean Distress Scores for Persons Living Either Within 5 Miles of TMI or Between 41 and 55 Miles Away from TMI at Four Time Periods*

| | DISTRESS MEASURES | | | | |
|---|---|---|---|---|---|
| Distance from TMI (miles) | Upset about TMI[1] | TMI a Serious Threat[2] | Behavioral Symptoms[3] | Somatic Symptoms[4] | Attributed Symptoms to TMI[5] |
| April 1979*** | | | | | |
| Within 5 | 3.36 | 3.10 | .30 | .18 | .69 |
| 41-55 | 2.21 | 2.13 | .07 | .05 | .29 |
| Difference | 1.15** | .97** | .23** | .13** | .40** |
| July 1979 | | | | | |
| Within 5 | + | 3.16 | .32 | .37 | + |
| 41-55 | + | 2.46 | .17 | .21 | + |
| Difference | | .70** | .15** | .16** | |
| January 1980 | | | | | |
| Within 5 | 2.59 | 2.39 | .40 | .51 | .25 |
| 41-55 | 1.95 | 1.60 | .18 | .35 | .04 |
| Difference | .54** | .79** | .22** | .16** | .21** |
| October 1980 | | | | | |
| Within 5 | 2.29 | 2.46 | .40 | .40 | .27 |
| 41-55 | 2.13 | 1.82 | .39 | .38 | .12 |
| Difference | .16 | .64** | .01 | .02 | .15* |

Effects of age, sex, education, marital status and income have been controlled.

1. How upset are/were you about TMI? Scale 1-5; 5 — Very upset, 1 — Not at all upset.
2. How serious a threat is/was TMI to family's safety? Scale 1-4; 4 — Very serious, 1 — Not at all serious.
3. Do/Did you have one or more of the following symptoms in the past 2 weeks (or, in the case of April, in the 2 weeks of the crisis): lack of appetite, overeating, sleeplessness, shakes, trouble thinking, irritability, or anger? 1 — Yes, 0 — No.
4. Do/Did you have any one or more of the following symptoms in the past 2 weeks (or, in the case of April, in the 2 weeks of the crisis): stomachaches, headaches, diarrhea, frequent urination, rash, abdominal pain, sweating spells? 1 — Yes, 0 — No.
5. Do you think your symptoms are due to TMI? 1 — Yes, 0 — No. (Only for persons reporting symptoms)

* Indicates that the difference between groups is significant at the $p < .05$ level.

** Indicates that the difference between groups is significant at the $p < .01$ level.

\+ July 1979, data is not available for upset and attribution of symptoms.

*** Data for April, 1979 were collected in the July, 1979 survey and, therefore are retrospective.

over time for both the close and far groups though, as with ratings of upset, the drop was sharper for respondents within 5 miles of TMI. The differences between close and far groups were still statistically significant in October, 1980, eighteen months after the original crisis.

4. The attribution of symptoms to Three Mile Island, which involved only respondents who reported behavioral or somatic symptoms, dropped sharply for

both groups during the 15 months studied, but, as with perception of threat, attribution was still significantly higher close to TMI in October, 1980.
The overall pattern of these findings shows reductions in distress for both the close and far groups over the 18 month period studied. There was a sharper rate of reduction for the group close to TMI than for far away.

While the data shown in Table 1 indicate long term distress, other studies of mental distress showed only short term stress following the crisis. Studies by Dohrenwend et al for the Presidential Commission on Three Mile Island (2) utilized the "demoralization" scale of psychological distress in several studies beginning immediately after the crisis and extending into May, two months later. They found demoralization scores markedly raised immediately after the accident, but found that they dropped sharply in April and May. In the May sample responses of mothers close to Three Mile Island were compared with responses of mothers in a control group approximately 100 miles away and it was found that they were not significantly different from each other. Eight months later a Penn State/Department of Health survey (7) found similar results utilizing the Langner scale of psychological distress which correlates very highly (over .90) with the demoralization scale. They found, in January, 1980, that there was no significant difference in scores on the Langner scale for persons close to Three Mile Island compared to persons in a control group from 41 to 55 miles away.

On the surface it appears that studies utilizing the Demoralization and Langner scales contradict results discussed above using symptom checklists and attitude questions. However, a careful examination of the two sets of data indicate two reasons for their differences. First, the Langner and Demoralization scales include many symptoms not included in the Somatic or Behavioral symptom lists. These symptoms characterize severe mental distress, e.g., "bothered by feelings of sadness or depression," "wondering if anything is worthwhile anymore," "feared going crazy," "felt completely helpless," etc.

A second reason for the different results is in the severity of symptoms assessed. For example, the "somatic" symptom scale asked if the respondent had experienced a headache in the past two weeks while the Langner scale asked if the respondent had experienced headaches "often." Since the Langner and the somatic symptom scales were asked in the same survey (January, 1980) it was possible to compare these two ways of asking about the same symptoms. When respondents were asked if they had had a headache in the past two weeks, there was a significant difference between persons within 5 miles of TMI and persons 41 to 55 miles away. However, when the same respondents were asked if they experienced headaches often, there was no difference between the two groups. These comparisons of the two sets of scales suggests that one set (the behavioral and somatic scales) assessed responses to normal day to day stresses while the second set (Langner and Demoralization scales) were more concerned with debilitating symptoms characteristic of mental patients. This suggests that the long term stress that resulted from the TMI crisis was in the "normal" rather than "pathological" range. This conclusion is consistent with findings by Bromet discussed below.

*Long term stress effects on special groups*

The studies cited above were primarily concerned with the total population in the vicinity of Three Mile Island. Several other studies focussed on special groups thought to be especially sensitive to the crisis. These groups included mothers of young children, workers at the TMI facility, and mental patients.

Bromet (1) carried out an extensive study of 312 mothers in the TMI area and compared their stress levels with those of 124 mothers in Beaver County, Pennsylvania, approximately 200 miles away, but also in the vicinity of another Nuclear Power Facility. The purpose of this design was to isolate the effects of the Three Mile Island accident from the effects of living near a nuclear facility. Bromet collected data approximately 9 months and 12 months after the crisis. A number of standardized instruments for assessing psychological distress were utilized by Bromet, as well as a complex measure of social support. Results of her analyses were as follows. 1) TMI mothers reported significantly more mental distress than did the control group mothers at both interviews (December, 1979 and April, 1980). 2) Mothers with a history of psychiatric disorder reported higher frequency rates than did comparable mothers in the control group, but onl in the April, 1980, survey. 3) While the symptom levels for TMI mothers at both 9 and 12 months were higher than for the control group, they were in the "normal" rather than "pathological" range. This finding is consistent with findings of the Penn State-Department of Health studies discussed earlier.

Kasl, Chisholm and Eskenazi conducted a study of workers at the TMI facility and compared their responses to workers at another nuclear facility approximately 40 miles away (12). They found that, compared to the control group, TMI workers perceived more danger to their health, had lower job satisfaction, expressed more uncertainty about their job future and were less likely to want to see their son or daughter work for their company. However, interestingly, they found that TMI workers were less willing to accept criticism of their company than were control workers. The authors concluded that TMI workers were showing ambivalence about their place of employment. Bromet (1) carried out a similar study with the same TMI workers several months later and compared their responses to comparable workers at another nuclear power plant 200 miles from TMI. Bromet reported higher levels of anxiety and depression among workers at the facility compared to control group workers. However, Bromet points out that these differences should be interpreted cautiously since rates for the TMI workers were higher than for the control group even before the crisis in March of 1979. An interesting additional finding in Bromet's study was that TMI workers felt more rewarded by their jobs than did control group workers, possibly because of the challenge of the special problems at the TMI facility.

A third population studied intensively by Bromet was clients of mental health out-patient facilities in the vicinity of TMI compared to similar patients near a nuclear facility 200 miles away. The findings were generally negative with little difference in symptom levels between the two groups.

In conclusion, a number of studies dealing with stress among persons in the vicinity of Three Mile Island showed that stress levels were high during the crisis, but dropped shortly afterwards to levels that were greater than for comparable control groups, but, nonetheless, were within a "normal," i.e., nonpathological range. Finally, as time has passed the number of distressed persons has consistently decreased to where, in October, 1980, 18 months after the crisis, several of the distress measures which had been greater close to TMI than for a control group, no longer showed statistically significant differences.

## SUMMARY

There is a remarkable degree of consistency among different studies of the social, psychological, and economic effects of the Three Mile Island nuclear accident in March, 1979. These studies generally agree that the major impact of the crisis was felt during the few weeks immediately following the accident and that, while some effects persisted over the following year, long term effects were of a low magnitude.

Temporary evacuation, which lasted less than two weeks for most evacuees, caused short term economic loss of approximately $90 million to individuals and local communities. It was during the two week evacuation period that psychological stress was the greatest, whether one evacuated or not. After the immediate crisis was over, concern about the situation at Three Mile Island persisted, maintained, in part, by continuous attention in the news media. This concern was evident in attitudes as well as reports of stress-related somatic and behavioral symptoms. However, these were within the "normal" rather than "pathological" range.

Concern about TMI was sufficiently high shortly after the crisis to affect people's stated intentions to move permanently out of the area, but not sufficiently high to significantly effect residential mobility behavior in the year following the crisis. Similarly, while symptom reporting was elevated among persons living near TMI, little evidence of increased demand for medical services could be found. Distress levels gradually fell over an 18 month period following the accident to the point where symptom reporting was no longer higher close to TMI as compared to further away. However, widespread concern about what has and may happen at Three Mile Island still exists among persons living close to the facility, as indicated by attitudes toward TMI and belief in rumors about alleged negative health effects.

## REFERENCES

1. Bromet, E. *Three Mile Island: Mental Health Findings,* Western Psychiatric Institute and Clinic, 1980.

2. Dohrenwend, B., Dohrenwend, B., Kasl, S. and Warheit, G. *Report of the Task Force on Behavioral Effects of the President's Commission on the Accident at Three Mile Island,* 1979.
3. Flynn, C. *Three Mile Island Telephone Survey, Preliminary Report on Procedures and Findings,* U.S. Nuclear Regulatory Commission, Washington, D.C., 1979.
4. Gamble, H. and Downing, R. *Effects of the Accident at Three Mile Island On Residential Property Values and Sales,* U.S. Nuclear Regulatory Commission, Washington, D.C., 1981.
5. Goldhaber, M., Houts, P., and DiSabella, R. *Mobility of the Population Within 5 Miles of Three Mile Island During the Period from August, 1979 Through July, 1980.* Report submitted to the TMI Advisory Panel on Health Research Studies of the Pennsylvania Department of Health, 1981.
6. Governor's Office of Policy Planning, *The Socio-Economic Impacts of the Three Mile Island Accident: Final Report,* Harrisburg, PA: 1980.
7. Houts, P., Miller, R., Tokuhata, G., Ham, K. *Health-Related Behavioral Impact of the Three Mile Island Nuclear Incident Parts I and II.* Report submitted to the TMI Advisory Panel on Health Research Studies of the Pennsylvania Department of Health, 1980.
8. Houts, P., Miller, R., Ham, K., and Tokuhata, G. Extent and duration of psychological distress of persons in the vicinity of Three Mile Island *Proceedings of the Pennsylvania Academy of Science* 54:22-28, 1980.
9. Houts, P., DiSabella, R., and Goldhaber, M. *Health-Related Behavioral Impact of the Three Mile Island Nuclear Incident Part III.* Report submitted to the TMI Advisory Panel on Health Research Studies of the Pennsylvania Department of Health, 1981.
10. Houts, P., Henderson, R., and Miller, R. *Family Practice Use and Response to the Three Mile Island Crisis.* Paper presented at the North American Primary Care Research Group, Lake Tahoe, CA, 1981.
11. Hu, T., Slaysman, K. Ham, K., and Yoder, M. *Health-Related Economic Costs of the Three Mile Island Accident.* Report submitted to the TMI Advisory Panel on Health Research Studies of the Pennsylvania Department of Health, 1981.
12. Kasl, S., Chisholm, R., and Eskenazi, B. The Impact of the Accident at the Three Mile Island on the Behavior and Well-being of Nuclear Workers: *Report of the Task Force on Behavioral Effects of the President's Commission on the Accident at Three Mile Island,* 1979.
13. Scranton, W. *Report to the Governor's Commission on Three Mile Island.* Harrisburg, PA, 1980.
14. Shearer, D. *Three Mile Island Nuclear Accident Community Impact Study on Real Estate.* Harrisburg, PA: Greater Harrisburg Board of Realtors, 1980.

*Chapter Fifteen*

# Problems in the Manufacturing Industry in High-Technology Areas

**D.E. Zappa**
President
VECTOR CORPORATION
3700 Butler Street
Pittsburgh, Pa. 15201

D.E. Zappa is President of Vector Corporation, Pittsburgh, Pennsylvania, a company engaged in the machining of naval and commercial nuclear components. He is also President of D.E. Zappa Associates, Inc., Consulting and Design Engineers, Pittsburgh.

In a highly technologically-oriented society, there are demands placed upon industry to increase productivity via a myriad of modalities, while maintaining the integrity of the customers' requirements. These very technological demands create problems for management in the areas of machine optimization, tooling approaches and quality assurance, plus the H-Factor. These problems affect both economics and energy consumption in the manufacturing of highly-technological products.

The critical concerns which I will address in this presentation are applicable to any manufacturing facility, regardless of size.

## MACHINE OPTIMIZATION

The age of numerically controlled equipment (which I will hereafter refer to as NC equipment) has been both beneficial and, conversely, detrimental to many manufacturing cycles. In setting critical path items, the rationale of leaning towards NC equipment optimizations versus conventional equipment optimizations has not been completely justified, and has frequently resulted in higher unit costs and a wasteful expenditure of net energy. For example, in the manufacturing of a complex component, there are certain manufacturing modes that mandate the use of NC equipment, due to the intracacies of the procedures that must be employed in the manufacturing sequence and the quantity of the components being produced. It, therefore, becomes essential that certain NC procedures be employed. Moreover, the optimization of NC equipment in this manufacturing modality proves to be very cost effective. After qualifying operations of the NC program, this virtually insures a product that eliminates costly in-process inspections, guaranteeing (in a normal operational mode) an acceptable component. However, there are also operations, such as loose tolerance hole drilling, being performed on NC equipment that should be performed on conventional-type equipment, in order to optimize this mode in the manufacturing operation. The aforementioned example depicts under-utilization of the conventional equipment and over-utilization of NC equipment that should be geared to the manufacture of more sophisticated components. Unfortunately, this practice creates a problem in both areas: first, it utilizes the NC equipment time frame on an operation that could be readily moved and assigned to a conventional-type machine; second, it inhibits the scheduling of NC equipment for the manufacture of more complex parts. Further, it is not cost effective to utilize this more sophisticated NC equipment for such menial operations. In order to optimize both NC equipment and conventional-type equipment, customer requirements for any particular time frame dictate the judicious utilization of the specific piece of equipment that will augment and interface with ancillary equipment in order to meet a schedular requirement. This not only effects production, but, concomitantly, is economically sound, because there is no additional utilization of energy outside this time frame.

## TOOLING APPROACHES

Our attention is immediately drawn to *cost-effective* tooling. We are currently in the age of over-utilizing carbide tooling. There are many operations performed using carbide tooling, i.e., drilling, turning and boring, that could readily be performed with high-speed steel tooling. For example, when carbide tooling is programmed into either NC equipment or conventional-type operations unnecessarily, the impact on the cost of manufacturing the customer's product can be as much as 20% to 30% of the selling price. This is incredible! Let us address the reasons for this. First of all, carbide tooling is not a panacea. Illogically, the thinking is that when manufacturing operations involve carbon steel, or any mild steel product for that matter, carbide-type tooling is a must. Actually, the manufacturing operation of these materials can be accomplished by utilizing high-speed steel (h.s.s.) tooling at a fraction of the cost. Moreover, when a company is oriented to the use of carbide tooling, inventories of this type are very *cost ineffective* when compared with stocking h.s.s. tooling. Unfortunately, this syndrome is prevalent in our highly-technological society of today. For example, a 1¾ " diameter carbide insert-type drill with required attachments costs $395.00; a 1¾ " diameter h.s.s. twist drill costs $115.00. Obviously the price ration is 3 to 1. We can see, then, that this approach to tooling requirements (carbide versus h.s.s.) grossly affects our competitive position in the world marketplace. This is not to mitigate against the proper utilization of conventional-type tooling.

## QUALITY ASSURANCE

End product quality is now being subjected to international criticism and, therefore, it is incumbent upon the manufacturing community to stress that a high level of workmanship be maintained. The demands of our society mandate the use of sophisticated commensuration devices. This adds another dimension in exactness, record keeping, traceability and accuracy in all phases of the manufacturing cycle. This obviously adds costs to the end product. Industry is thus faced with the dilemma of complying with customer demands and, at the same time, meeting the challenge of remaining competitive while burdened with these added constraints. Sophisticated on-line sample plan inspection must be implemented without affecting production. To assure the quality of these highly complex components, it is imperative that critical flow path operations are checked in the correct time frame to prevent any interruption of the manufacturing sequence, while still maintaining the quality of the end product. If a company is to remain competitive, without sacrificing profits, it is essential that it incorporate a well-planned program of on-line sample plan inspection, interfaced with a correct time frame in the production cycle. A concomitant benefit of such a program is the reduction of scrap, waste and component rework time. Conversely, a poorly planned program can add substantially to net costs.

## THE H-FACTOR

The H-Factor, or more specifically, the human factor, is a problem in the high-technology manufacturing industry which is grossly overlooked.

Chronic changing of personnel from machine to machine, especially in the area of manufacturing complex components, decidedly adds to the overall cost of the component. The reasons are manifold, but to cite a few—employee apathy, reduced morale and stress.

Employee Apathy — This is a result of many underlying issues, but primarily, management's inconsistency in dealing with employee needs. For example, management has been very quick to reprimand and discipline an employee who has not performed to company expectations. On the other hand, management is parsimonious in recognizing employee excellence and/or positive participation in overall company philosophy.

Reduced Morale — This can be ameliorated by proper machine assignment, participation by the employee in company/management decisions, and rewarding continued quality of productivity for any given manufactured component. Too often management overlooks the valued participation by an employee in the manufacturing of the end product. This can cause a protraction of work, thus leading to lower production, extended energy use due to a time frame change, a higher cost per unit and even poor quality assurance.

Stress — This can be either machine-induced or peer-induced, due to the success or failure syndrome. The environment of most manufacturing facilities is a complex issue. With continued demands for quality end products, the employee-operator is chronically subjected to optimal performance requirements and frequently is unable to maintain the imposed standards. The net result is a physiological stress induction which directly affects the employee's performance. Coupled with this is the machine-induced phenomenon. Computerized equipment constantly functions at a pace and a level at which an employee cannot function. This will enhance stress in the operator. To circumvent this machine-induced stress, an appropriate time frame as far as operator assignment to any given machine, without affecting production, must be carefully scrutinized and integrated.

Another contributing factor is boredom induced by the operation of NC equipment, because the skilled operator has been reduced to a mere button pusher. The oneness of pride in his work is reduced proportionately to the repetitiveness of the product being produced. It is, therefore, essential that, from time to time, the operator is reminded of the fact that he has the capability of recognizing any problem in the machining sequence by utilizing three principles — praising the employee, empathizing with employee problems and encouraging the employee to have pride in his work, thus adding a new dimension in a positive direction to the H-Factor.

In this paper I have called attention to four target areas. Obviously, we can splinter these taxons, but the purpose of this paper is to address the more comprehensive issues in the manufacturing industry in high-technology areas.

*Chapter Sixteen*

# The Ecological and Economic Impact of Anthracite Mining in Pennsylvania

**Donald L. Miller, Ph.D.**
**and**
**Richard E. Sharpless, Ph.D.**
Department of History
LAFAYETTE COLLEGE
Easton, Pa. 18042

Dr. Donald L. Miller, chairman of the American Civilization Program at Lafayette College, received his Ph.D. from the University of Maryland. He is author of *The New American Radicalism,* a study of left-wing politics and thought in the era of the Great Depression, and of numerous articles and reviews on modern American history. Miller is currently writing two books: a biography of Lewis Mumford for Charles Scribner's Sons; and, with Richard E. Sharpless, a popular history of the anthracite industry and culture (University of Pennsylvania Press). In 1980-81 he received a Fellowship from the National Endowment for the Humanities to begin his Mumford book. He is a recent recipient of Lafayette's Jones Award for excellence in teaching and research. With Professor Sharpless he organized the Northeastern Pennsylvania Regional Studies program at Lafayette.

Dr. Richard E. Sharpless, Associate Professor of History at Lafayette College, Easton, Pa., received a Ph.D. from Rutgers University. A specialist in Latin America, he is the author of *Gaitan of Colombia* and articles on the modernization process in history. He currently is writing, in collaboration with Dr. Donald L. Miller of Lafayette College, a popular history of the anthracite industry and culture. With Professor Miller he organized the Northeastern Pennsylvania Regional Studies program at Lafayette, which focuses on the industrialization process in the Lehigh Valley and anthracite region of Pennsylvania.

At the opening of this century, coal was King in Northeastern Pennsylvania. Within an area of less than 500 square miles, hard coal—anthracite—gave rise to one of America's first raw capitalist industries, with its powerful coal barons, stupendous technological achievements, and bloody labor struggles. For over 125 years a single industry—coal—dominated this region, and here the industrializing process assumed a most nakedly brutal form. In less than a generation an unspoiled wilderness of steeply pitched mountains and rolling valleys was made over into a dreary wasteland of culm banks, coal breakers and ramshackle company towns. By the time of the Civil War, only three decades after the opening of the anthracite fields, the region's forests had been stripped completely of their timber stands and its streams and rivers dangerously polluted by raw sewage from coal settlements and acid seepage from the mines.

The industry also extracted a terrible human cost. Deep in the coal seams, men and boys from Wales, Ireland and Poland worked in total darkness, often knee-deep in water, at the most dangerous job of the day. Accidents were a common occurrence, almost a way of life, and few miners past the age of 40 failed to contract respiratory disease from inhaling the dust, fumes, and foul air. No other early American industry inflicted more heedless destruction on man and the environment than anthracite mining.[1]

Yet the shiny "black diamonds" these capitalists and miners pulled from the earth broke this country's dependence upon foreign coal and triggered an industrial revolution that made the United States a world power.[2] At the turn of the century, Pennsylvania's clean-burning anthracite was the Northeast's chief domestic heating fuel. By 1917 this great industry, now a sprawling industrial combination ruled by the nation's largest investment houses, produced 100 million tons of coal, and employed a workforce of nearly 180,000 men.

That was more than a half-century ago. Today, with production down to less than 6 million tons a year, and a workforce of barely 3,000, anthracite accounts for only one percent of the nation's coal supply and about seven percent of Pennsylvania's production.[3] The region's thinly populated towns, abandoned mine shafts and decaying coal breakers are poignant reminders of a once dominant industry and way of life.

But is this the final episode in the saga of anthracite? Oil and natural gas—the fuels which supplanted it—are becoming scarcer and ever more expensive. And the nation's anxious search for alternative energy sources has sparked a revival of interest in hard coal as a regional fuel. As a result, the mood in the anthracite region today is guardedly optimistic. There is a greater demand for anthracite, particularly as a home heating fuel, but producers, uncertain of the future markets, are as yet unwilling to undertake the necessary capital investments to increase production significantly.

Where does anthracite stand today? And what is its immediate future? This essay addresses these questions through a brief survey of the industry's growth, power, and decline. It is our belief that a close understanding of anthracite's history must inform every aspect of the policy debate over its future. No program for its revitalization can afford to ignore the history of its past uses and misuses. As anthracite looks to the future, it cannot escape its own past.

Anthracite is found in abundance in only one part of the world: an area of Northeastern Pennsylvania little more than 120 miles long and 40 miles wide. Although some anthracite is found in several other states, and in a scattering of Asian and European countries, three-quarters of the earth's hard coal deposits are located in Pennsylvania.

The state's anthracite region is divided into four geological sections which, in turn, comprise three major trade divisions or fields: the Wyoming, the Lehigh and the Schuylkill.

The Wyoming, or Northern, field, shaped like a long canoe, extends from Forest City in Susquehanna county 55 miles west to Shickshinny on the Susquehanna river. Though little more than six miles wide, this field has the deepest deposits, and its coal the highest carbon content. Scranton and Wilkes-Barre, its commercial centers, are the region's largest cities. At the peak of production in 1917 one-half of Pennsylvania's anthracite was shipped from the Wyoming trade area.

The Lehigh, or middle field, is the smallest of the three. Covering an area of approximately 51 square miles, it includes the eastern middle geological division centered around Hazleton, and the Southern basin to the east of Tamaqua. The coal here lies in steeply pitched beds and is difficult to mine. This area produced one-sixth of the region's total output around 1900.

The largest, and southernmost, of the fields is the Schuylkill. Within its 238 square miles are two geological divisions: the western middle section, extending from Carbon County to Columbia and Northumberland counties on the Susquehanna; and the Southern basin, west of Tamaqua, an area almost entirely in Schuylkill county. This field produced about one-third of all anthracite, but the coal was often "bony" and mixed with sandstone. Today it holds the greatest reserves. The largest city in the Schuylkill field is Pottsville, the "Gibbsville" of John O'Hara's stories, whose native American character sets it apart from the neighboring towns of Shamokin, Mount Carmel and Shenandoah—tight little ethnic communities which evoke images of Eastern Europe with their onion-domed churches and Slavic neighborhoods.[4]

All of Pennsylvania's coal—anthracite and bituminous—was formed at roughly the same time, more than 250 million years ago, when the state was a flat, moist plain, filled with steamy swamps thick with tall trees and wide-spreading ferns. As these giant plants decayed and died, they fell to the bottom of the swamps, gradually forming a spongy, brown vegetable matter called peat. In time, this peat was squeezed down and sealed off from the air by mud, sand, and new debris carried in by the periodic eastward movement of the large inland sea that covered the center of the continent. This increased pressure combined with heat from the earth's core to transform the peat, over millions of years, into bituminous, or soft, coal.

In the western part of the state the bituminous coal is deposited relatively close to the surface in flat-lying beds. But in Northeastern Pennsylvania the Appalachian Revolution heaved up the flat plain and folded the earth's crust, creating the mountains of the region. Here the additional pressure of the earth's upheaval produced coal of unusually high carbon content.

The average fixed carbon content of anthracite is 86%, with volatile matter of only 4.3%. This makes it more difficult to ignite, but it burns longer and cleaner than bituminous, making it more attractive for domestic use. It is also easy to store and is more resistant to deterioration.

Yet the very geological pressures which produced this high-quality coal also restrict its wider use. Anthracite lies in severely pitched veins, which often reach below the area's water table, making it more difficult and dangerous to mine than bituminous.[5] This gives bituminous a significant cost and safety advantage over anthracite in today's energy market. Anthracite, however, was the first widely used American coal for domestic and industrial purposes. Its history, from early discovery and exploitation by small, risk-taking capitalists to monopoly domination of the industry by the big railroad companies and New York banking houses, replicates the major developments of the American industrializing process.

No one knows for certain who discovered anthracite. Each locality has its own story. But its first commercial uses were by local blacksmiths in the Wyoming Valley, who mined it from outcroppings and burned it in their forges. Anthracite, however, was not seen as a valuable regional fuel until early in the 19th century, when this country experienced its first energy crisis. As the forests near growing cities like Philadelphia were cut away, the cost of wood and charcoal shot upward. The war of 1812 also interrupted the flow of British "sea coal" into the country, leading a number of Eastern businessmen to invest in anthracite coal lands.

Technological improvements further contributed to the acceptability of the new fuel. Anthracite was difficult to ignite and farmers generally preferred wood in their fireplaces, while iron plantations used charcoal. In 1810, however, Judge Jesse Fell of Wilkes-Barre introduced an open, natural draft grate, which could be inexpensively installed in fireplaces; and in 1833 Frederick W. Geisenheimer took out a patent for smelting iron with anthracite in blast furnaces. These discoveries helped to incite national interest in anthracite as a domestic and industrial fuel.[6]

Transportation technology was the key to the first coal boom. Early mining, largely a pick and shovel job, was a relatively simple matter of working the outcroppings on the side of mountains. Getting the coal to market was another matter. There were no easily accessible roads and the region's treacherous rivers, watery deathtraps for the first primitive coal arks, ruined many an early investor's dreams of instant wealth.

Only with the coming of the coal canal, an innovation borrowed from Britain, did anthracite mining grow into a major industry. Beginning in the 1820s, enterprising merchants and capitalists constructed a series of canals that joined the Schuylkill, Lehigh, and Wyoming Valleys with the major tidewater ports of New York and Philadelphia. These anthracite canals, with their river links, formed the nation's first comprehensive inland transportation system. They were among the most heavily capitalized private corporations in the nation prior to the 1850s. The names of the men who directed these companies reads like a Who's Who of American men of finance in those years. Two of the largest anthracite canal companies, the Delaware and Hudson and the Lehigh Coal and Navigation, were also, by 1850, the two largest mining companies in the region.[7] From this time until the industry's decline, those companies which controlled the region's transportation network controlled the mining and marketing of its coal.

Anthracite production increased steadily throughout the 19th century. In 1820 only 365 tons were produced, but by 1850 production exceeded three million tons, making anthracite one of the fastest growing industries in the country. Towns grew spectacularly and villages appeared overnight. Port Carbon, with only one family in 1829, had 912 inhabitants a year later. Pottsville, a hamlet of a few houses and taverns in 1825, was a bustling trading town of 7,500 people by 1850.[8]

By this time railroads were competing with canals for control of the coal trade. In 1855 the Philadelphia and Reading Railroad was shipping twice as much coal to Philadelphia as the Schuylkill canal. It was not long before the railroads completely supplanted the canals as the major coal transporters. In the process, they transformed the character and financial structure of the coal industry.

The anthracite industry evolved in the classic capitalist pattern, from small enterprises operated by individual entrepreneurs, through family-owned companies, to giant corporations organized into trusts. In the early years of the industry, thousands of small entrepreneurs leased or bought the small tracts they worked and shipped their products to market over the primitive canal and river systems. But as the demand for coal increased, these small producers, particularly in the Schuylkill region, were driven out by the big canal and railroad companies. The pick and shovel capitalist was replaced by the company that could afford water pumps, ventilation fans, breakers, rolling stock, and other technological advances. Even the canal barons could not stop this inexorable movement toward industrial combination and coordinated corporate control. In the years following the Civil War, the railroads and the large banking houses gained complete hegemony in the anthracite industry. By 1896 ninety percent of the industry was controlled by five of the railroad companies serving the anthracite fields. As late as 1936 ten companies controlled three-quarters of the anthracite production.[9]

In 1917 anthracite production reached a peak of 100 million tons, but markets had shifted significantly. This low sulfuric coal remained the preferred heating fuel of the industrial Northeast, but it had been replaced by bituminous, which was cheaper to mine and transport, as the principal fuel for smelting iron. In 1910 anthracite produced only two percent of the nation's pig iron. And in the decades following World War I it lost its monopoly over the domestic fuel market to oil, natural gas, and, eventually, electricity.[10]

The beginning of the real decline began after the long strike of 1925-26. From these years on, except for a brief period of prosperity during World War II, production dropped steadily. Other factors, apart from oil and gas, contributed to the decline. Strict immigration laws reduced the influx of foreign workers willing to accept lower wages, and unionization stabilized the labor market to some extent. Also, fewer men were willing to risk the dangers of an occupation that had killed 16,000 miners between 1900 and 1930. Capital, too, went elsewhere as increased taxation of coal companies reduced incentives for investment. Another blow was delivered when diesel engines were introduced on railroads, eliminating an important anthracite market.[11]

Today, after more than 150 years of mining, there are approximately 17 billion tons of anthracite still in the ground, of which perhaps 7 to 8 billion tons are economically recoverable with available technology. Despite this significant reserve, anthracite presently accounts for only one percent of all coal mined in the United States. In 1980 5.5 million tons were produced, and this was the first year in recent history in which production actually increased over the previous year.[12]

Distribution of anthracite within the United States in 1978 totalled approximately 3.9 million short tons, or 77.3 percent of production. Of this amount, 1.6 million short tons, or 41 percent, went to residential and commercial users, the largest traditional market for the product. Electric utilities took another 27.3 percent, or slightly more than one million short tons. Industrial users consumed 748 thousand short tons, or 19.2 percent, and the iron and steel industry 476 thousand short tons, or 12.2 percent. Mining operations used 0.3 percent.[13]

Anthracite exports in 1978 totalled 1.14 million short tons, or 22.7 percent of total production. The principal consumers (88.6 percent of total exports) were Canada (405 thousand short tons), South Korea (331 thousand short tons), and U.S. military forces in West Germany (276 thousand short tons).[14] Though total tonnage exported to U.S. military forces has been declining steadily over the past decade, it still is considered one of the most secure markets of the industry.

While anthracite historically was extracted primarily from underground mines, today that method accounts for only six percent of production. Strip mining provides 51 percent of total output, with processing of culm and silt banks, remnants of past mining, yielding 42 percent. About one percent is obtained by dredging. In 1978 there were 92 strip mines, 80 underground mines, 112 culm bank recovery operations, and 6 river dredges working.[15] The total value of all hard coal produced in that year was approximately $177.6 million.[16]

The center of production also has shifted. Presently Schuylkill county, the area of largest reserve, provides almost half of all anthracite. Luzerne county, in the Northern field, accounts for about one-fourth, with the other anthracite counties far behind. Most future production is expected to remain in the Southern and Western Middle fields.[17]

At its peak the anthracite industry was dominated by companies organized into trusts associated with the J.P. Morgan interests. Absentee ownership was centered in New York and Philadelphia. As the industry declined, land and equipment passed into the hands of small operator-owners. In 1978 38 companies were carrying out mining operations in the region, and 15 of them accounted for 71 percent of total production.[18] None of the companies is large by contemporary standards. For example, the Jeddo-Highland Coal Company, the leading producer, had 350-400 employees in 1974 and a payroll of about $3 million. These companies have been able to stay in business primarily because of their adoption of labor-saving technology and sophisticated mining techniques.[19]

By 1978 there were only 3,474 miners working on an average day in the anthracite fields, and they averaged 223 working days per year. Average production per miner per year was 1,451 short tons, which represented a continued long-term productivity decline (6.6 percent below the previous year, 1977).[20]

Thus, from its position as the major source of employment in the region early in this century, mining has declined until today the industry employs less than two percent of the labor force. The sharpest drop in employment occurred in the decades immediately following the Second World War. As recently as the late 1940s there were 80,000 workers still employed in anthracite. The major factors involved in the post-war decline were the continuing shift to other energy sources for residential and commercial heating, even within the region itself, and the growing importance of capital intensive strip mining.

The legacy of anthracite's decline has been long-term economic depression, high unemployment, and heavy out-migration of the younger members of the work force. Employment levels are well below the national average: in the mid-1970s, 73.8 percent compared to a national average of 79.2 percent for males, and 42.2 percent versus 42.8 percent nationally for females.[21] In the expansive 1950s unemployment in the region never fell below 10 percent, and as late as 1962 unemployment compensation provided the major single source of income in the region's largest cities, Wilkes-Barre and Scranton.[22] Women have fared better than men during much of the post-war era because they have found employment in the textile industry, which now is the major single source of income in the region. This, however, created severe cultural and psychological shocks as proud former miners became dependent upon their wives. As might be expected, medium family incomes in all anthracite counties except Dauphin, where the state capital is located, are below the Pennsylvania average.

Beginning in the early 1960s concerted efforts were made at local, state and federal levels to bring about a revitalization through economic diversification.

The city of Hazleton, for example, largely through the efforts of a local organization of business and civic leaders, achieved a notable economic comeback, including high employment, by the mid-1970s. Scranton and Wilkes-Barre —the latter especially hard-hit by the Susquehanna river flood of 1972—made notable economic advances. In addition to textiles, other major industries now located in the region include construction, primary and fabricated metals, electrical equipment, and food products.

But despite limited successes in attracting industry and diversifying the economy, the region remains one of the most economically depressed areas in Pennsylvania and, indeed, of the Northeastern United States. And it is unlikely that coal, in the short term at least, will bring about a general economic revival. Though there might be substantial increases in anthracite production, new capital-intensive open pit mining methods preclude the employment of a large labor force, while prospects for expanding underground mining are uncertain, at best.

One of the most important obstacles to increased underground mining is the lack of experienced, skilled workers. The decline of deep mining has substantially reduced the available pool of underground miners. Most experienced men remaining in the region are near or past retirement age. Younger workers are not attracted by deep mining. They prefer to seek other types of employment, or leave the region entirely. Since the high production era of the 1920s, the populations of most mining towns have declined by over 50 percent.[23] Recent efforts to attract underground miners to the region and to train new ones have not been successful.

The anthracite industry not only left a crippled economic base, it also inflicted severe environmental damage. It is estimated that one-quarter of the region's 484 square miles has been disturbed by mining operations. Refuse banks remain from the days of underground mining. With the spread of strip mining in the 1920s the mountains of the region were blasted and ripped apart to reach the coal seams, leaving the land barren and almost uninhabitable. Today, at least 800 culm banks still despoil the landscape, along with scores of abandoned collieries and deteriorating mine patch towns.

Perhaps the most disastrous environmental impact was upon the surface and sub-surface water systems. Refuse from underground mines and stripping operations destroyed or altered the courses of streams. The vast underground network of interconnected mines, driven beneath the water table, required drainage tunnels and constant pumping which disturbed the existing hydrologic systems. After abandonment, these deep mines filled with water, and the acidic overflow polluted—and continues to pollute—surface waters in the region and beyond. According to Pennsylvania environmental officials, acid mine drainage accounts for 76 percent of the damage to the state's streams. A more recent, but ominous, development, has been the illegal dumping of chemical wastes into abandoned mine openings. Eventually, the wastes find their way into the streams and rivers of the region. The most affected river has been the Susquehanna, where extremely dangerous toxic wastes from mine drainage tunnels have been found. Chemicals like dichlorobenzene have been identified in drinking water of downstream towns. Since the Susquehanna flows into the Chesapeake, the potential for widespread ecological and human damage is great.[24]

Another problem is subsidence, which almost always follows anthracite mining. Throughout the region, highways and roads have been twisted and damaged, underground utilities frequently broken, and the foundatins and walls of buildings cracked. Houses, automobiles and, occasionally, people, have fallen into holes caused by subsidence.[25]

The degree of subsidence depends upon the amount of coal extracted, the number, depth, and thickness of the coal seams, and the thickness and strength of the rock strata between the coal beds. The decline of underground mining has greatly increased the occurrence of subsidence in the Northern field. As the abandoned mines fill with water, the submerged soft and brittle rock formations lose their bearing strength. One result of this process was experienced by a section of Wilkes-Barre following the hurricane of 1972, when flooding mines caused extensive cave-ins and property damage.[26]

In addition to the flooding of abandoned mines, removal of the supports in the underground tunnels contributes to subsidence. Timbers in idle mines eventually rot and collapse. The "robbing" of support pillars of coal during second and third mining often results in cave-ins on the surface.

Mine fires also have created dangerous conditions in the region. Presently, there are eight fires in abandoned underground mines, eight culm bank fires, and four landfill fires still burning.[27]. A fire that began in 1962 under the Columbia county town of Centralia is weakening coal support pillars in old mine tunnels and gangways, raising the possibility of a general community evacuation. In February, 1981 a 12-year old boy nearly lost his life when the ground in his back yard collapsed beneath him. The greater immediate danger, however, is posed by carbon monoxide fumes from the mine fire which have found their way into a number of Centralia homes.[28]

Until well into the 20th century there was almost no concern for the environment on the part of coal companies or government agencies. The attitude of *laissez faire* predominated. Little attention was given to reclamation. As a result, though blame can be laid upon the mining industry for past environmental damage, responsibility for rectification cannot be placed upon companies which in most cases no longer exist.

Presently, safeguards exist for protecting the environment in the region, and the mining companies themselves recognize that past practices are unacceptable. Indeed, the revitalization of anthracite mining will have positive effects upon the environment. Regulations require the purification of contaminated water in new mines, and the contouring of land in surface operations. Second and third mining over previously worked areas provides, under existing legislation, for land reclamation at the cost of the operators. Deep pit mining, possibly down to 1000 feet or more, eventually would eliminate acid mine water drainage by dealing with the problem of underground mine water pools.[29]

The benefits of re-mining already have been demonstrated in Pennsylvania. During the decade 1967-77, in both the bituminous and anthracite fields, about 30,000 acres of previously worked land was reclaimed. If public agencies had accomplished this, the estimated cost to the taxpayer would have been nearly $100 million.

Over the past decade public awareness of environmental issues has increased substantially throughout the region. Disasters such as the Centralia mine fire serve to increase that awareness. It is unlikely that, even given the revitalization of anthracite, practices like those that ravaged the environment in the past will be condoned. State and local environmental groups like the Sierra Club, the Schuylkill Preservation Alliance, and the Hegins Township Environmental Association educate the public about the costs and benefits of mining. Further, government agencies with a previous record of opposing expansion of coal use, such as the U.S. Environmental Protection Agency, now support its increased use because of the benefits it could have for land reclamation. Anthracite's image among environmentalists clearly is changing.[31]

But what of hard coal's future? What are the prospects for the revitalization of the industry and, hence, the entire regional economy?

The prospects for the near future are, at best, mixed. Rather than a spectacular "boom," the industry is likely to experience a slow, steady growth during the remaining decades of this century.[32] Anthracite will remain primarily a regional fuel. Naturally, unforeseen developments affecting the international energy configuration, such as a severe and sustained interruption in the flow of overseas oil to domestic markets, could alter the outlook for anthracite. But in the absence of this, dramatic increases in production are unlikely. Expansion of output probably will occur at or near the rate of 11.7 percent experienced between 1979 and 1980.[33]

There are several factors limiting the growth of the industry. Though hard coal mining continues to be dominated by a handful of large producers, they themselves are small when compared with companies in the soft coal fields. The largest anthracite companies, for example, employ no more than several hundred workers. They generally are undercapitalized. Without assured markets for expanded production, markets that have been uncertain in any case, financing for increased production is difficult to obtain. Further, in the absence of expanding demand, operators are reluctant to invest millions in the expensive technology necessary today. They are still faced with the geological difficulties of mining anthracite, which makes it more expensive to extract than its competitor, bituminous.

Anthracite operators are unsure of their markets. This was demonstrated most recently during the winter of 1980-81, when they refused to expand production to meet a sudden surge in demand in New England. In another much publicized market, South Korea, misunderstandings and conflicts over costs and quality, as well as unstable political conditions in that Asian country, created uncertainty. Thus, miners and consumers view each other suspiciously, waiting for the other party to act.[34]

There are other constraints, as well. So long as there is no coherent national energy policy, in which anthracite has an assured place, and which includes programs designed to enhance hard coal's attractiveness, producers will be reluctant to invest. Such a policy logically would include mandatory requirements for oil burner conversion to coal, reconstruction of the rail transport system, and expanded, modernized port facilities. But it also would have to include investment and tax incentives for the industry, and a more rational, coordinated federal-state approach to regulations. The present chaotic situation regarding environmental and safety requirements actually works as a dis-incentive that has hobbled production and forced some small operators out of business.

Another drawback is the lack of funding for research and development of new mining techniques and technology. The sharply curving pitch and faulting of the coal seams, especially in the Southern field, the area of greatest reserve, makes it impossible to adopt the mechanical equipment of the bituminous industry. If the unit labor cost of anthracite is to be reduced, new technology must be developed.[35]

Yet the short-term outlook is not entirely bleak. Production increases of 10-11 percent annually indicate that the industry is not stagnant or dying, but growing, if only slowly. Existing established markets are unlikely to disappear, given the uncertainty over energy sources. Some limited efforts, such as the growing use of anthracite stoves in New England as a supplemental heat source, mandated use of hard coal in new and renovated state buildings, and improved port facilities for export, should assure some additional growth.

The prospects for significantly expanded production in the 1980s and 1990s depend upon highly uncertain factors. As stated above, the cost and availability of other energy sources, mostly oil, are critical. Should the cost of home heating oil and gas continue to inflate rapidly, the possibilities for conversion to clean burning anthracite, ideal for residential and commercial heating, are enhanced. Additionally, the use of anthracite in the manufacture of synthetic fuels, already utilized on a small scale in Pennsylvania, would be encouraged. Similarly, industrial uses of anthracite, such as gasification and coking, could be expected to increase as the cost of soft coal increases. While foreign markets remain uncertain, rising world costs of oil would provide incentives for increased use of anthracite, from fresh coal to reprocessed culm and silt.[36] Presently, a French corporation is investing more than $10 million in a deep mine joint venture with an American firm in western Schuylkill county, despite prevailing wisdom that deep mining is too costly.[37]

Perhaps the most important recent development regarding an anthracite comeback was a decision by the federal government that exempted hard coal from the requirement that coal burning power plants be equipped with scrubbers designed to remove sulfur from smoke. Since scrubbers cost hundreds of millions of dollars, clean burning, low sulfur anthracite becomes competitive with bituminous in parts of the Northeast.

Largely as a result of this decision, a consortium of power-generating companies—Pennsylvania Power and Light, Philadelphia Electric, and Allegheny Electric—are discussing the possibility of constructing a mine-mouth generating complex of three to six 500-megawatt units in Schuylkill county. Such a facility would provide several hundred construction, approximately 100 mining, and over 100 operation and maintenance jobs. Each 500-megawatt unit would consume one million tons of anthracite annually. The giant bituminous producer, Consolidation Coal Company, has been invited to participate as the mine operator. Such an immense project, however, if it does finally reach beyond the study stage, would require 10 years lead time, and thus would not be operational until the 1990s. But it could result in a tripling of production.[38]

It is in the area of power generation that the outlook for anthracite is brightest. Projections indicate that the United States will require more than one billion tons of coal annually for electrical generation by the end of the century.[39] If the present environmental restrictions on bituminous hold, power companies in the Northeast will find anthracite increasingly attractive.

It is doubtful that coal will be king again in the anthracite region of Pennsylvania. If the prognosis for the industry is slightly more hopeful than a decade ago, there still is little likelihood that an anthracite boom will occur. Nor will even a sharp increase in anthracite production power a regional economic revival. Though coal made this region, and will play an important part in its immediate future, a general economic revival will occur only through the continued diversification of its economy.

## NOTES

1. The best general history of the anthracite industry is still Peter Roberts, *The Anthracite Coal Industry: A Study of Economic Conditions and Relations of the Co-operative Forces in the Department of the Anthracite Coal Industry in Pennsylvania* (New York: Macmillan, 1901).
2. Alfred D. Chandler, Jr., "Anthracite Coal and the Beginnings of the Industrial Revolution in the United States," *Business History Review* 46 (Summer, 1972): 141-81; Eliot Jones, *The Anthracite Coal Combination in the United States* (Cambridge: Harvard University Press, 1914).
3. Anthracite Task Force, *1977 Report* (Washington, D.C., United States Department of Energy, 1977).
4. Victor R. Green, *The Slavic Community on Strike: Immigrant Labor in*

*Pennsylvania Anthracite* (Notre Dame, Indiana: University of Notre Dame Press, 1968).

5. William E. Edmunds and Edwin F. Koppe, *Coal in Pennsylvania* (Harrisburg: Pennsylvania Geological Survey, 1968), 1-18.
6. Erksine Hazard, "History of the Introduction of Anthracite Coal into Pennsylvania and a Letter from Jesse Fell, Esq. of Wilkes-Barre on the History and Use of Anthracite into the Valley of Wyoming," *Memoirs of the Historical Society of Pennsylvania,* 2:155-164; H. Benjamin Powell, *Philadelphia's First Fuel Crisis: Jacob Cist and the Developing Market for Pennsylvania Anthracite* (University Park: Pennsylvania State University Press, 1978); John N. Hoffman, "Anthracite in the Lehigh Valley of Pennsylvania, 1825-1845," Contributions From the Museum of History and Technology, Smithsonian Institution, *Bulletin* 252 (Washington, D.C.; Smithsonian Institution Press, 1968); Frederick M. Binder, *Coal Age Empire: Pennsylvania Coal and Its Utilization to 1860* (Harrisburg: Pennsylvania Historical and Museum Commission, 1974).
7. H. Benjamin Powell, "The Pennsylvania Anthracite Industry, 1769-1976," *Pennsylvania History,* XLVII, no. 1 (January, 1980): 6-10; Chester Lloyd Jones, *The Economic History of the Anthracite-Tidewater Canals* (Philadelphia: University of Pennsylvania Press, 1908); H. Benjamin Powell, "Schuylkill Coal Trade, 1825, 1842," *Historical Review of Berks County,* 39 (Winter, 1972-1973): 14-17, 30-34; Ronald L. Filippelli, "The Schuylkill Navigation Company and Its Role in the Development of the Anthracite Coal Trade and Schuylkill County 1815-1845" (M.A. Thesis, The Pennsylvania State University, 1966); The Delaware and Hudson Company, *A Century of Progress: History of the Delaware and Hudson Company 1823-1923* (Albany: J.B. Lyon, 1925).
8. Ronald L. Filippelli, "Pottsville: Boom Town," *Historical Review of Berks County,* 35 (Autumn 1970): 126-129, 155-157; C.K. Yearly, Jr., *Enterprise and Anthracite: Economics and Democracy in Schuylkill County, 1820-1875* (Baltimore: The Johns Hopkins Press, 1961); D.F. Shafer, *A Quantitative Description and Analysis of the Growth of Pennsylvania Anthracite Coal Industry, 1820-1865* (New York: Arno Press, 1977).
9. Jules I. Bogen, *The Anthracite Railroads: A Study in American Railroad Enterprise* (New York: The Ronald Press Company, 1927); Marvin W. Schlegal, *Ruler of the Reading: The Life of Franklin B. Gowen, 1836-1889* (Harrisburg: Archives Publishing Company, 1947); Dan Rose, *Energy Transition and the Local Community: A Theory of Society Applied to Hazleton, Pennsylvania* (Philadelphia: University of Pennsylvania Press, 1981); Scott Nearing, *Anthracite: An Instance of a Natural Resource Monopoly* (Philadelphia: John C. Winston Co., 1915).
10. Powell, "Pennsylvania Anthracite," 20.

11. There is an excellent analysis of the reasons for the industry's decline in Rose, *Energy Transition,* passim; See also M. J. Shapp and E.J. Jurkat, *New Growth . . . New Jobs for Pennsylvania* (Philadelphia: The Shapp Foundation, 1962); Anthracite Coal Commission, *1938 Report of the Anthracite Coal Commission* (Harrisburg: Commonwealth of Pennsylvania, 1938); Anthracite TaskForce, *1977 Report;* T. Bakerman, "Anthracite Coal: A Study in Advanced Industrial Decline," Ph.D. Dissertation, University of Pennsylvania, 1956.
12. Dr. Jerry Pell, "Anthracite: What is Its Future in the Northeast?" Statement presented by the Director, Division of Anthracite, Office of Fossil Energy, U.S. Department of Energy, before the Northeast-Midwest Congressional Coalition, U.S. House of Representatives, Pittsburgh, PA, April 22, 1981, p. 17.
13. U.S. Department of Energy, *Coal—Pennsylvania Anthracite 1978: Energy Data Report* (Washington, D.C.: U.S. Department of Energy, Energy Information Administration, 1980), p. viii.
14. Ibid.
15. Dr. Jerry Pell, "The Prospect for Anthracite as a National Energy Resource," Presented by the Director, Division of Anthracite, Resource Applications, U.S. Department of Energy, at the 6th Energy Technology Conference, Washington, D.C., February 12, 1979, p.3.
16. *Coal—Pennsylvania Anthracite 1978,* p.3.
17. The locations of anthracite resources and current amounts of strippable reserve, by fields, are: Northern, 2 billion tons (400 million tons strippable); Eastern Middle, 0.3 billion tons (43 million tons strippable); Western Middle, 3 billion tons (225 million tons strippable); Southern, 12 billion tons (1 billion tons strippable). Information provided by Dr. Jerry Pell, Director, Division of Anthracite, U.S. Department of Energy.
18. Quote in Pell, "The Prospect for Anthracite as a National Energy Resource," p. 3, from: "How Major Producers View Appalachian Potential," *Keystone Coal Industry Manual* (New York: McGraw Hill, 1975).
19. Dan Rose, *Energy Transition and the Local Community: A Theory Applied to Hazleton, Pennsylvania* (Philadelphia, University of Pennsylvania Press, 1981), pp. 169, 171.
20. *Coal—Pennsylvania Anthracite 1978,* p.2.
21. *Anthracite Task Force Report,* Section II, p. 6.
22. Rose, *Energy Transition and the Local Community,* p. 169.
23. Powell, "The Pennsylvania Anthracite Industry," p. 26.
24. Rose, *Energy Transition and the Local Community,* pp. 171, 175.
25. Laurie Loken, "Flooding, Acid Mine Water Drainage, and Subsidence in the Northern Field of the Anthracite Region," paper prepared for the Northeastern Pennsylvania Regional Studies Seminar, Lafayette College, Easton, PA, May 1981.

26. William M. Griffith and Carl J. Romanelli, eds., *The Wrath of Agnes* (Wilkes-Barre, PA, Media Affiliates, 1972), p. 112.
27. Paul E. Carpenter, "Anthracite and the Languishing Boom," supplement to the *Pottsville Republican,* Pottsville, PA, March 20, 1981, p. 36.
28. Teresa Carpenter, "Burn, Centralia, Burn," *Village Voice,* May 20-26, 1981.
29. Statement of Karl F. Goos to the Energy and Mineral Resources Sub-Committee of the Energy and Natural Resources Committee, U.S. Senate, Washington, D.C., March 24, 1981.
30. Testimony of Walter N. Heine, Director, Office of Surface Mining Reclamation and Enforcement, before the Sub-Committee on Energy Development and Applications, Committee on Science and Technology, U.S. House of Representatives, Washington, D.C., September 4, 1980.
31. Carpenter, "Anthracite and the Languishing Boom," pp. 19, 36.
32. Ibid, p. 39.
33. Ibid, p. 3.
34. Ibid, pp. 37, 38.
35. Powell, "The Pennsylvania Anthracite Industry," p. 26.
36. Pell, "Anthracite: What is its Future in the Northeast?"
37. Carpenter, "Anthracite and the Languishing Boom," p. 24.
38. Ibid., pp. 5-8; Pell, "Anthracite: What is its Future in the Northeast?"
39. Carpenter, "Anthracite and the Languishing Boom," p. 4.

*Chapter Seventeen*

# Energy from Non-Conventional Sources

**Heikki K. Elo, P.E.**
LEHIGH FORMING CO., INC.
P.O. Box 799
Easton, Pa. 18042

Mr. Heikki K. Elo is president of the Lehigh Forming Co., Inc., which specializes in the design and construction of solid waste resource recovery facilities. He graduated from Worcester Polytechnic Institute, Worcester, Mass. with a degree in engineering. Mr. Elo is a member of the Pennsylvania and National Society of Professional Engineers and the Sigma Xi Research Fraternity as well as a charter member of the Resource Recovery Committee of the American Society for Testing and Materials.

The search for fuels to satisfy the world's energy needs has led to the development of fuels from non-conventional sources. These alternative fuels are unique because they are renewable: solar, biomass, wind and geothermal.

Alternative fuels development has progressed rapidly during the past ten years due to governmental incentives, decreasing supplies of traditional fuels, and increasing energy needs. During the next decade, alternative fuels will become a significant source of energy in the United States. This chapter will examine the various types of alternative fuels and their potential impact on the United States' energy needs.

The goal of alternative fuels production is to supplement, rather than replace, traditional fuels. By using alternative fuels where possible, we can reduce our dependence on foreign oil, release oil for use in areas where the substitution of other fuels is not possible, and help to keep fuel costs as low as possible.

It has become evident that energy independence is a goal which the United States should strive for. It is too dangerous politically and economically to be dependent on uncertain sources for an essential commodity. Alternative fuels can be used in place of imported oil and gas to reduce or eliminate our dependence upon them.

By their very nature, many fuel uses, transportation being the largest, preclude the substitution of alternative fuels for petroleum based products. However, alternative fuels can be substituted for oil in many instances, thereby releasing that oil for use in transportation and related industries.

We cannot realistically expect a technological breakthrough of the magnitude which would solve all our energy problems; the safest approach to U.S.A. and global energy needs is to develop as many potential fuel sources as possible. Alternative fuels will not be able to replace traditional fuels, but they can make a significant impact on U.S.A energy needs within the next decade.

The present energy crisis, which began in 1973-74, has brought about dramatic changes in our lifestyles and in the global political climate. We must make every effort to conserve traditional energy sources, and to develop as many alternative sources as possible.

Energy pervades every phase of our lives: it is required for:

- Transportaion — automobiles, trains, trucks, boats, airplanes, space vehicles
- Industrial machinery — drills, lathes, welders, grinders, shears, punches, presses
- Basic metals industries — foundries, rolling mills, furnaces
- Boilers — steam for chemical reactions, electrical pump, fans, mills
- Mining — explosives, crushers, conveyors, screens
- Food — fertilizers, pesticides, combines, milling, pelletizing, cooking
- Forestry — debarking, saw mills, conversion, chipping, pulping
- Fabrics — forming of natural and man-made fibers into threads, weaving, conversion of cloth into clothes and industrial products

• Comforts — heating, cooling, lighting, humidifying

The energy for these, and many other, needs comes from a variety of fuel sources. Total U.S.A. energy needs are approximately 80 quads per year.* One barrel of oil at $36/bbl. provides 5.8 million BTU. U.S.A. energy needs (80 quads) are equivalent to 13,790 million bbl. of oil at a cost of approximately $500 billion per year. This figure represents about one-fourth of the U.S.A. gross national product, $2.1 trillion.

Most of the commodities which can be used for energy are also used in other areas. Oil is not only our most important energy source, it is also a raw material for chemicals, plastics, and fertilizer. Natural gas is refined into other gases, chemicals, and products. Coal is the backbone of the synthetics, chemical, and drug industries. The use of wood as a fuel source is negligible when compared to its use in other areas — lumber, paper, cardboard, and forest products. The competing uses for fuels are significant, and drain a portion of them from the energy market. These commodities are available for energy use only if energy users can pay a competitive price for them.

All of the factors we have discussed: U.S.A. dependence on foreign oil, the need for petroleum based products in specific uses, and competition for fuels in the marketplace, encourage development of alternative fuels.

Our energy needs are satisfied by fuels from a variety of sources: traditional fuels — domestic crude oil, imported crude oil, domestic natural gas, imported natural gas and coal; and alternative fuels—nuclear, hydroelectric, tidal, wood, peat, propane, butane, wind, geothermal, solar and biomass. The following is an assessment of the potential for those fuels meeting our energy needs in the future.

## DOMESTIC CRUDE OIL

Production of domestic crude oil pumped from the ground will gradually decrease as it has been for the past several years. This decrease will be offset by production of oil from shale and tar sands; and, in fact, domestic oil production will experience a slight overall increase during this century due to the employment of new technologies such as deep drilling, steam flushing, and offshore drilling.

Rising world oil market prices will encourage development of the more exotic technologies, which have hitherto been economically unfeasible. The extent to which these new methods are employed will be controlled by the price consumers are willing to pay.

*one quad = 1,000,000,000,000,000 British Thermal Units

## IMPORTED CRUDE OIL

In our analysis we have totally discounted imported crude oil as an energy source. They have done this for two reasons, 1) to demonstrate that energy independence can be achieved, and 2) because supplies of imported crude are subject to instantaneous interruption and/or sky-rocketing prices. We cannot be assured that the OPEC countries will always be willing to supply us with oil, or at what price that oil will be supplied. For these reasons, long range energy planning should assume the worst scenario—that the United States will not import any crude oil.

## DOMESTIC NATURAL GAS

Domestic production of natural gas will increase. Previous pricing regulations and the lack of adequate pipelines severely limited growth in this area, and contributed to waste in the form of flashing.

The availability of easy to reach natural gas will decrease due to the reduction in oil pumping. However, the author feels that the currently existing legal limitations pertaining to the use of natural gas from coal mines and similar operations will be removed. Removal of these limitations will release a significant amount of natural gas for fuel use.

## IMPORTED NATURAL GAS

As with imported crude oil, we have assumed future imports of natural gas to be nonexistent.

## COAL

Today, coal is an underpriced form of energy. As its popularity increases prices will rise, thereby encouraging development of previously uneconomical coal fields. Severe environmental regulations on coal mining and use will be reduced, and the high cost of other fuels will encourage development of better antipollution devices for the coal industry.

New techniques such as the addition of limestone to the fuel will reduce $SO_2$ and $SO_3$ emissions. Calcium sulfate in the ash will increase, but air quality will be improved significantly.

The principal energy uses of coal will be in steam electric generating plants. These steam plants, using large quantities of coal, will be equipped with the necessary air pollution control equipment. The other major users of coal will be the cement and lime industries.

## NUCLEAR

This analysis predicts a slight increase in nuclear energy production. Although there are many nuclear generating stations proposed, public opposition will be a major hindrance to their construction. Many of the nuclear plants which have been proposed, or are under construction will never be completed due to public opposition and the logistical difficulties involved in preparing plans for the large scale evacuation of communities in the event of an accident.

## HYDROELECTRIC

Nearly all of the hydroelectric potential in the United States has been utilized. The few potential dam sites which do remain are the subject of fierce opposition by local residents and environmentalists. Even if we were to build those additional dams, their impact on total U.S.A. energy needs would be insignificant.

The areas where hydroelectric power has its greatest potential are in the relatively undeveloped regions of the world: Africa, Asia, and South America. Development of hydroelectric plants in these continents would reduce world dependence on oil, and would provide relatively affordable energy for developing nations.

Despite the potential for growth in hydroelectric use in third world countries, its overall proportion of the world energy balance is gradually decreasing due to increased energy use. By the year 2000, hydroelectric power will supply about 3% of the world's energy needs.[1]

## TIDAL, WAVE, AND OCEAN TEMPERATURE DIFFERENCES

The harnessing of tidal and wave power is limited to a few locations in the world. Although the possibility of capturing wave and tidal energy has been proven experimentally, it has never been tried commercially. "Only high tides under certain conditions may be regarded as promising for large-scale energy generation."[1] There is a tremendous amount of energy in the ocean, but the need for technological breakthroughs and the huge expenditures for research will keep tidal and wave power from becoming a significant energy source during the remainder of this century.

The temperature difference between the surface and the depths of the ocean is a potential source of energy for which very little research has been done. In tropical regions this temperature difference can range from 25-30 °C. The greatest forseeable drawback to this source of energy is the potentially negative environmental effects it may have on marine life due to large-scale changes in ocean temperature. The small amount of research which has been done to date, and the potential negative environmental impact make this field one which will have little effect on the world energy balance.

## WIND

Wind, although it contains a great deal of energy, represents a limited source of containable power. The use of wind as an energy source should be developed to its fullest, but its large-scale potential is not that great.

The primary drawback to harnessing wind power is its ethereal nature. There are very few locations in the world which offer substantial and steady flows of air. Because of its inconsistency, it will probably only be used in small scale installations which have back-up energy sources.

## GEOTHERMAL

As with tidal power and hydroelectric power, geothermal energy is extremely site specific. Sites for construction of geothermal power stations are scarce; they are usually in zones of young volcanic activity.

There are three major sources of geothermal energy: steam, hot water, and hot, dry rock. Underground steam is the scarcest of the three and has already been harnessed in most of the locations where it is available.

Hot water averages only about 100°C and has a high mineral content. The relatively low temperature of the water means that an extremely large amount of water must be pumped out of the ground. The removal of these enormous reserves could severely damage the surface structure of the earth unless cooled water is pumped back underground.

Needless to say, pumping large volumes of water with high mineral content out of the ground, through heat exchangers, and back into the ground will require virtually continuous replacement of worn parts. However, the rising costs of energy in general will encourage development of these facilities despite their high overhead.

The largest reserves of geothermal energy are in the form of hot, dry rock which is either naturally porous or artificially cracked. The technology needed to gather the available energy has not yet been developed at a commercial level. The largest problem associated with this particular resource is the development of economically feasible methods for recovery of the energy which lies from 3000m-5000m below the surface of the earth.

The maximum worldwide yearly potential for all three forms of geothermal energy has been estimated at about 0.5% of the current world energy balance.[1] This is equivalent to about 1.02 quads per year.

## SOLAR/BIOMASS

Solar/biomass energy represents the greatest alternative source of energy available presently and in the near future. "Biomass refers to all products of photosynthesis,"[2] including wood, corn, algae, and solid waste materials derived from the products of photosynthesis. Because of the wide range of fuels covered in this area we have broken it down into several categories: solar, direct and indirect; municipal waste; industrial waste; and other forms of biomass.

## SOLAR

Solar energy is a virtually unlimited resource; approximately 5000 times more energy than we could use falls on the earth each year[1]. As with all alternative fuels technologies, the key to success lies in the development of economical technologies to capture the available energy. For example, development of film photoelectric cells which cost several hundred times less per surface area than existing semiconductor cells.[1]

To date, large-scale solar facilities have been impractical because the power of a facility is in direct proportion to the size of the plant, and because of the daily and seasonal fluctuations in solar radiation.

Considering the tremendous potential of solar energy, every effort should be made to develop economical large and small-scale solar technologies. The long range potential of solar energy staggers the imagination; it is not only available in vast quantities, but the fuel itself is free of cost and non -polluting. The development of solar technologies can be compared to the field of electronics; given the proper recognition and incentives, technology will make a quantum leap.

## MUNICIPAL WASTE

Waste to energy systems are the fastest growing renewable energy sources available today. Rapid growth in this field can be attributed to the simultaneously decreasing supplies of traditional fuels at increasing prices, and the decrease in affordable landfill space. Production and use of refuse derived fuel (RDF) reduces the need for landfills and provides a significant alternative source of energy.

The United States Department of Energy has estimated that our nation produced approximately 154 million tons of municipal refuse during 1978. The projected U.S.A. refuse production for the year 1990 is 200 million tons[2-4] Conversion of this refuse into fuel would make a significant impact on national energy production. The 75% of refuse which is combustible will represent 2.1 quads of energy in 1990.

Waste is defined by the federal government as "garbage, refuse and other discarded materials, including solid waste materials from industrial, commercial, and agricultural operations, and from community activities.[3]" Municipal waste consists of the following fractions: 75% combustibles, 6% ferrous metals, 19% heavy non-combustibles (including glass, ceramics, non-ferrous metals, etc.).

A wide range of systems are presently available for conversion of the combustible fraction of the waste into energy:

1. incineration — enclosed burning of raw waste.
2. pyrolysis — chemical decomposition of waste by heating it in an oxygen-deficient atmosphere — material is broken down into various hydrocarbon gases and carbon-like residue.
3. composting — natural conversion of organic material to humus by micro-organisms.
4. hydrolysis — cellulosic material is chemically processed to yield a diluted sugar solution which can be made into ethyl alcohol.
5. methane production — an odorless, colorless, low BTU, flammable gas formed by the decomposition of organic waste.
6. refuse derived fuels (RDF)
   a. shredded RDF — waste is processed by separating out the combustible fraction and then shredding it until it resembles mattress ticking — fluffy fuel is burned in boilers to produce steam.
   b. densified RDF — waste is processed in a way similar to shredded RDF, then it is subsequently densified (fuel pellets) and burned in boilers to produce steam.

One of the drawbacks with most forms of municipal waste to energy systems has been their low BTU/cf ratios, the fuels are so voluminous that they cannot be transported and are so difficult to handle that they are not readily acceptable by fuel users. The development of fuel pellets (densified RDF) has changed this, and created a revolution in this field. For example, a silo 30 feet in diameter and 100 feet high can hold 16.3 BBTU* of pellets. It would take a building 100 ft. wide x 50 ft. high x 200 ft. to contain the same number of BTUs of shredded RDF.

In fuel pellet production, the three fractions of the waste stream are mechanically separated. Ferrous metals are recycled; heavy, non-combustibles, which represent 10% by volume of the total waste stream, are landfilled; and the remaining light combustible fraction is extruded through a rotary die to produce dense, fuel pellets with uniform characteristics.

The average unit density of fuel pellets, which is about 30 pcf, makes them a desirable fuel source for several reasons. Fuel pellets are easily conveyed, they can be stored in and metered from conventional silos or bunkers, and they can be transported in trucks or by rail easily and cheaply.

Their unique cylindrical shape prevents the pellets from bridging in silos or on

---

*BBTU = Billion British Thermal Units

conveyors. This shape allows air to flow easily through a mass of pellets so that 1) heat buildup is reduced during storage, and 2) air flow through a boiler fuel bed is uniform raising their combustion efficiency.

The BTU/lb. ratio of fuel pellets is comparable to that of coal — municipal fuel pellets contain about 7500 BTU/lb. The ratio of BTUs per cubic foot for pellets (230,000 BTU/cf) is nearly ten times that of shredded RDF (24,000 BTU/cf).

Pellets can be burned either as a primary fuel source or as a supplement to coal. Their shape, size, and density make fuel pellets very compatible with coal. They can be used in existing coal-capable conveying, storage, transportation, and combustion systems with no modifications to those systems. Because fuel pellets contain a nil amount of sulfur, they can be burned in conjunction with high sulfur coal while remaining within emissions standards.

One of the greatest advantages of pelletized fuel over other forms of RDF is that it can be burned in any solid fuel boiler with no modifications to the boiler. Their versatility is evident from the list of combustors in which they can be burned: pulverized coal-fired steam generators, stoker fired generators (ram, travelling grate, steam jet, or vibrating grate type), fire tube boilers, fluidized bed boilers, kilns, ovens, and roasters. They can be burned to produce steam, or can be used in a cogeneration cycle to produce both steam and electricity.

Refuse derived fuel pellets are an immediately available, inexpensive, alternate source of energy. They burn cleanly and are highly compatible with coal and coal capable equipment. Conversion of this nation's municipal waste into fuel pellets could provide 2% of this nation's energy needs by the year 2000.

## INDUSTRIAL WASTE

There are a large number of industrial plants which generate combustible waste. In many cases the advantages of waste to fuel systems are even greater for industry than they are for municipalities, due to the high costs of waste disposal for private industry and the unique opportunity to burn waste produced at a given plant in their own boilers.

Some of the most common industries for whom waste to energy systems are practical are:

- assembly plants — packaging waste from parts shipped to assembly site.
- paper industry — waste from paper conversion plants, printing plants, packaging plants, etc.
- any industry which generates petroleum-based combustible sludges—auto, furniture, printing, and chemical plants, etc.

Refuse derived fuel pellets are particularly attractive to private industry because they have many of the same handling and burning characteristics of coal. The compatability of these two fuels eliminates the need for expensive plant

TABLE I

*Energy Sources in the U.S.: Present & Future*

| | QUADS PER YEAR CURRENT | 2000 | POTENTIAL |
|---|---|---|---|
| Domestic Crude Oil | 23.0 | 25.0 | |
| Imported Crude Oil | 14.0 | — | |
| Domestic Natural Gas | 17.0 | 20.0 | |
| Imported Natural Gas | 3.0 | — | |
| Coal | 17.0 | 25.0 | 50.0 |
| Nuclear | 3.0 | 4.0 | 15.0 |
| Wood, Peat, Propane, Butane | 1.0 | 2.0 | |
| Hydro & Ocean | 2.0 | 2.5 | |
| Wind | — | 0.5 | |
| Geothermal | — | 0.5 | |
| Solar/Biomass | — | 5.0 | 25.0 |
| TOTAL | 80.0 | 84.5 | |

Clearly the largest contribution which alternative fuels can make during this century is as a supplement to traditional fuels. For the United States specifically, the use of alternative fuels will mean a reduction in our dependence on foreign oil, and a release of oil for use in those areas where the large-scale substitution of alternative fuels is not possible.

modifications and re-education of boiler room personnel.

The advantages for private industry to convert their waste into fuel are obvious. Waste disposal costs are reduced; energy costs are reduced; and, in some cases, hazardous waste disposal costs are eliminated.

## BIOMASS

This is another significant source of energy which has only recently been explored. It is based upon the dedicated growth of vegetation (e.g. trees, algae, crops) as fuel source. This vegetation may either be burned directly or processed into alcohol which will in turn be burned.

Wood, although increasing as a fuel source, will compete heavily in other areas —primarily paper and lumber. Its increasing value may lead to the use of branches and tree tops as fuel. The two largest disadvantages to the use of wood as fuel are its high moisture content, and the high costs of transportation from forest to energy.

Table I estimates the impact alternative fuels will make on the United States energy balance for the year 2000 in comparison to the present. We have arbitrarily set fuel imports at 0 quads to demonstrate that energy independence can be achieved if necessary.

The column labeled "potential" is for the long range impact of the various fuels. Solar/biomass fuels are unique because they are renewable; their use is not limited to finite reserves as with traditional fuels.

## REFERENCES

1. Hartnett, James, P., editor, *Alternative Energy Sources,* Academic Press, New York, 1976.
2. National Center for Resource Recovery, Inc., *Glossary of Solid Waste Management,* Washington, 1972.
3. National Center for Resource Recovery, Inc., *Let's Talk Trash,* Washington, 1973.
4. National Center for Resource Recovery, Inc., *Resource Recovery from Municipal Solid Waste,* Lexington Brooks, Lexington, Mass., 1974.
5. Reed, Tom and Bryant, Becky, *Densified Biomass: A New Form of Solid Fuel,* Solar Energy Research Institute, Golden, Colorado, 1978.

## PERTINENT LITERATURE

6. Alter, Harvey and Horowitz, Emanuel, *Resource Recovery & Utilization,* American Society for Testing and Materials, Philadelphia, 1975.
7. Committee for Economic Development, *Achieving Energy Independence,* Georgia Press, 1974.
8. Congressional Quarterly, Inc., *Continuing Energy Crisis in America,* Washington, 1975.
9. Duchesneau, Thomas D., *Competition in the U.S. Energy Industry,* Ballinger Publishing Co., Cambridge, Mass., 1975.
10. Elo, Heikki K., *Alternative Energy Research* (a series of papers, 1973-1981), Lehigh Forming Co., Inc., Easton, Pa.
11. Exxon Corporation, "The Offshore Search for Oil & Gas, fourth edition," New York, 1980.
12. Landsberg, Han H., Chairman of a Study Group, *Energy: The Next Twenty Years,* Ballinger Publishing Co., Cambridge, Mass. 1979.
13. Shell Oil Company, *The National Energy Outlook — 1980-1990,* Houston, 1980.

*Chapter Eighteen*

# Solar Energy — The State of the Art

**H. Maurice Carlson**
Professor Emeritus
of Mechnical Engineering
LAFAYETTE COLLEGE
Easton, Pa. 18042

Professor Carlson joined the faculty of Lafayette College in 1957. He attended Augustana College and earned bachelor's degrees in education and mechanical engineering at the University of Minnesota. He holds a master's degree in mechanical engineering from the University of Louisville and a master's degree in environmental science from Rutgers University. In addition, he did graduate work in nuclear engineering at Iowa State University. Professor Carlson served as Lafayette's head of the department of mechanical engineering for 21 years. From 1959 to 1962 he also served as director of engineering. During his career he has written a number of papers on energy related topics. Professor Carlson has lectured on various occasions and participated in numerous symposia over the years.

The oil embargo of 1973 marked the beginning of a new era; with it came the end of cheap abundant energy. It brought about a new realization of our energy dependence, and an increased awareness of the need to find alternative sources of energy to eventually replace those fossil energy sources, namely oil and gas, rapidly being depleted. One of the alternative renewable sources of energy receiving much attention is solar energy. "The amount of energy reaching the Earth's surface is so vast as to be almost incomprehensible. Two examples may make this clear. On a global scale, the solar energy that arrives in one to two weeks is equivalent to the fossil energy stored in all the Earth's known reserves of coal, oil, and natural gas. In the United States, the solar energy that reaches one-five hundredth of the country—an area smaller than that of Massachusetts—could, if converted at 20 percent efficiency, satisfy all our present needs for electrical power."[1]

It is the purpose of this paper to try to put into perspective the current "state of the art" in the present development and application of means to obtain energy from the sun in ways that would reduce our dependence on oil and natural gas and yet maintain a high standard of living and continued industrial growth, characteristic of the United States.

A casual look at current literature and publications might give the impression that seeking ways to harness the sun's energy is relatively new. This is not the case. In Ancient Greece, wood had become nearly non-existent as it had been extensively used as a fuel and as a building material for homes and ships. "As indigenous supplies dwindled and wood had to be imported, many city-states regulated the use of wood and charcoal. In the fourth century B.C., the Anthenians banned the use of olive wood for making charcoal. Most probably they passed this measure to protect their valuable groves from incursions by fuel-hungry citizens. Yet Athens' own supply lines stretched all the way across Asia Minor to the shores of the Black Sea. On the island of Cos, the government taxed wood used for domestic heating and cooking. Authorities on Delos, which had no indigenous supplies whatever, severely restricted the sale of charcoal. They believed that such a valuable energy source should not be controlled by a few powerful merchants—leaving the consumers to pay any price.

With wood scarce and the sources of supply so far away, fuel prices most likely rose. Fortunately, an alternative source of energy was available—the sun—whose energy was plentiful and free. In many areas of Greece, the use of solar energy to help heat homes was a positive response to the energy shortage. Living in a climate that was sunny almost year-round, the Greeks learned to build their houses to take advantage of the sun's rays during the moderately cool winters, and to avoid the sun's heat during the hot summers. Thus solar architecture—designing buildings to make optimal use of the sun—was born in the West.

Modern excavations of many Classical Greek cities show that solar architecture flourished throughout the area. Individual homes were oriented toward the southern horizon, and entire cities were planned to allow their citizens equal access to the winter sun. A solar-oriented home allowed its inhabitants to depend less on charcoal-burning braziers—conserving fuel and saving money."[2]

"Solar architecture and urban planning evolved along similar lines in ancient China. The streets of important cities were always laid out in orthogonal fashion aligned with the points of the compass. Whenever the site permitted, the preferred house plan bore a striking resemblance to the Olynthian homes of Greece in that its principal apartments were built on the north side of a courtyard that opened to the south. While there were few window openings on the north, east and west walls, large wood-lattice windows covered by translucent rice paper or silk were common on the south. Decorative overhangs protected the interior from the hot summer sun."[3]

In ancient Rome the great demand for wood to fuel industry, to build homes and ships, to heat public baths and the private villas created an energy shortage greater than that in Greece. "These local fuel shortages and the high cost of imported wood probably influenced the Romans to adopt Greek techniques of solar architecture. The Romands did more than copy the Greeks; however, they advanced solar technology by adapting home building design to different climates, using clear window coverings such as glass to enhance the effectiveness of solar heating, and expanding solar architecture to include greenhouses and huge public bath houses. Solar architecture became so much a part of Roman life that sun-rights guarantees were eventually enacted into Roman law."[4]

"From the days of Augustus in the first century A.D. until the Fall of Rome, the use of solar energy to heat residences, bathing areas and greenhouses was apparently fairly widespread throughout the Empire. Exactly how much the Romans relied on the sun is impossible to say, but the helicaminus or "solar furnace" room was common enough to provoke disputes over sun rights requiring adjudication in the Roman courts. As the population increased, buildings and other objects blocked the solar access of some heliocamini, and their owners sued. Ulpian, a jurist of the second century A.D., ruled in favor of the owners, declaring that a heliocaminus' access to sunshine could not be violated. His judgement was incorporated into the great Justinian Code of Law four centuries later:

> If any object is so placed as to take away the sunshine from a heliocaminus, it must be affirmed that this object creates a shadow in a place where sunshine is an absolute necessity. Thus it is in violation of the heliocaminus' right to the sun.

That his opinion was written into the Justinian Code in the sixth century strongly indicates that the construction of solar heated buildings continued until this late date.

Except for the community bath houses visited by the general public, solar heating was enjoyed by only the very rich in Rome. Mass use never caught on as it did in Greece. Rome did not plan for her poorer citizens. In sharp contrast to the

Greek spirit of democracy and social equality, the dominant Roman ideology favored class privilege—so that only the wealthy could build their homes with the proper solar orientation. Technologically, however, Rome made important advances in solar architecture—namely, transparent window coverings used as solar heat traps for homes, baths and greenhouses. And for the first time in recorded history, laws were set down establishing sun rights."[5]

History reveals that the Greeks, Romans, and Chinese knew of ways to use the sun's energy other than orienting buildings to take advantage of the sun's warming energy and using glass to create a heat trap. They "developed curved mirrors that could concentrate the sun's rays onto an object with enough intensity to make it burst into flames within seconds. Solar mirrors that could generate spectacular heat captured people's imaginations for centuries to come; they could be used constructively or possibly as terrible weapons of war.

The Greeks were the first Western people to describe "burning mirrors"—solar reflectors made of polished silver, copper or brass. The earliest burning mirrors probably consisted of many flat pieces of polished metal joined together to form a roughly curved surface. These were replaced by concave spherical mirrors dispensing with the bulkiness of many small flat-plane mirrors linked together . . .

The ancient Chinese also used burning mirrors primarily for religious purposes. Independently, they began making concave solar reflectors at about the same time as the Greeks. According to the Chou Li, a book of ceremonies written in approximately 20 A.D. which describes rituals dating far back into Chinese antiquity, "The Directors of Sun Fire have the duty of receiving, with a concave mirror, brilliant fire from the sun . . . in order to prepare brilliant torches for sacrifice."[6]

Down through the years research work continued in an effort to find better ways of capturing the sun's energy. "During the latter part of the sixteenth and throughout the seventeenth century, almost every scientist and gentleman experimenter investigated the curious powers of burning mirrors. Giovanni Magini, the Italian astronomer, wrote that he could melt "lead, silver or gold in small quantities such as a coin held firmly with tongs."[7]

Until the beginning of the nineteenth century the early uses of the sun's energy all centered around obtaining the energy in the form of heat. Warming of homes, heating of water for baths, and the use of mirrors to concentrate the sun's rays to melt metals and light fires were examples. The coming of the Industrial Revolution, bringing with it the use of machines that took the place of muscle-power of men and machines, resulted in the manufacture of goods on a scale far greater than at any other time in history. The resulting machines depended on the production of iron. By this time coal, as well as wood, was being used to supply the heat required in the founding of metals. Making a ton of iron required seven to ten tons of coal. Coal was the major source of heat for the recently developed steam engines and for heating the buildings and factories being built throughout

Europe. "France was at a disadvantage compared to other industrial countries because she had to import almost all of her coal. As a consequence, she lagged far behind rapidly industrializing England. So the French decided to pursue an aggressive program to step up domestic coal production. The plan worked, and output doubled in two decades—providing resources and power for iron smelters, textile plants, flour mills, and the many other new industries that began to appear by the second half of the nineteenth century.

Many Frenchmen now felt secure about the nation's energy situation, but not everyone shared this complacency. In 1860 Augustin Mouchot, a professor of mathematics at the Lycee de Tours, cautioned:

> One cannot help coming to the conclusion that it would be prudent and wise not to fall asleep regarding this quasi-security. Eventually industry will no longer find in Europe the resources to satisfy its prodigious expansion . . . Coal will undoubtedly be used up. What will industry do then?

Mouchot's answer was, "Reap the rays of the sun!" To show that solar power could be harnessed to run the machines of the Industrial Age, he embarked upon two decades of pioneering research."[8]

In the early years of his research, Mouchot was successful in producing several inventions including a solar oven, a solar still and a solar pump. 'Despite these successes, Mouchot had not yet attained his main goal: to drive a steam engine with sun power. The large volume of water inside the copper cauldron took a long time to boil, and the device produced steam too slowly to drive an industrial motor. Mouchot therefore substituted a one-inch diameter copper tube for the cauldron so that the smaller volume of water in the tube would heat much faster, generating steam more quickly. To collect the steam Mouchot soldered a metal tank to the top of the tube. The solar collector consisted of a parabolic, trough-shaped mirror that faced south and was tilted to receive maximum solar exposure.

Excitedly, Mouchot reported what happened when he connected this boiler to a specially designed engine:

> In the month of June, 1866, I saw it function marvelously after an hour of exposure to the sun. Its success exceeded our expectations, because the same solar receptor (i.e., reflector and boiler) was sufficient to run a second machine, which was much larger than the first.

Mouchot had invented the first steam engine to run on energy from the sun! He presented it to Napoleon III, who received it favorably."[9]

It was in 1860 when Mouchot, then 35, began his research and work with solar energy. After two decades of successful endeavor he returned to his field of mathematics. "But the time wasn't right for solar energy in France. The advent of better coal-mining techniques and an improved railroad system (most of France's coal lay at her borders) increased coal production and reduced fuel prices. In 1881 the government took one final look at the potential for commercial use of solar energy: it sponsored a year-long test of two solar motors—one designed by Mouchot and the othe by Pifre. The report concluded,

> In our temperate climate (France) the sun does not shine continuously enough to be able to use these devices practically. In very hot and dry climates, the possibility of their use depends on the difficulty of obtaining fuel and the cost and ease of transporting these solar devices.

Furthermore, the cost of constructing silver-plated mirrors and keeping them highly polished proved economically prohibitive for most uses. During ensuing years, however, the French Foreign Legion made some use of solar ovens in Africa. In remote areas of Algeria, people also used solar stills to obtain water, as one Paris correspondent reported:

> One of the great services that we owe to Mouchot's appliances is the distillation of brine water heavily charged with magnesium salt, which is abundant in the Africa desert. (His still) is a great benefit to settlers and explorers.

Although Mouchot did not succeed in bringing France into the "Sun Age," his pioneering efforts cross the threshold between scientific experimentation and the practical development of a revolutionary technology. He demonstrated a great variety of ways that solar energy could be used to benefit humankind and laid the foundations for future solar development."[10]

In America efforts were also being made to harness the sun's energy to produce mechanical power. About eight years after Mouchot had begun his experiments, John Erickson, a Swedish-American engineer and inventor, began work that resulted in his first solar powered steam engine in 1870. Two years later he produced a solar engine that used air as the working medium instead of steam. In 1888 Erickson had perfected his solar engine incorporating a more effective, less expensive reflector and was ready to enter into agreements with the Western farmers who were much in need of this engine for irrigation purposes. However, seven months later before any agreements for construction had been worked out, Erickson died at the age of 86. Because of earlier unfortunate experiences of losing inventions to others because of information he had revealed, he kept the details to himself and very little is known about the important aspects of his engine.

In 1902 the massive strike in the coal industry with its devastating effects emphasized the need for some alternative source of energy and many thought that solar energy was the most promising.

The most successful inventor to follow John Erickson was Aubrey Eneas, an engineer and inventor living in Massachusetts. In 1892 he began studying the possibilities of obtaining power from the sun and shortly thereafter he founded the Solar Motor Company of Boston with the express purpose of designing and manufacturing commercial solar engines. He was particularly interested in producing sun-powered pumps for irrigation in the Southwest part of the U.S. The market for these machines had great potential as conventional fuels were very expensive and the sun shone on seventy-five percent of the days. Eneas finished constructing his solar engine by 1899 and took it to Denver, Colorado where the direct rays of the sun were much hotter than in Massachusetts. After testing and making some changes to the supporting structure, he put the engine on public display at the only ostrich farm in the U.S. which had become an attraction for tourists from all parts of the nation. After two years of successful operation at the ostrich farm, Eneas incorporated in California and opened offices in Los Angeles. He sold several of the engines using the large truncated cone reflectors to concentrate the sun's rays. Damage to these reflectors through failure of the support structure in a wind storm in one instance and by a hailstorm in another convinced Eneas that engines using these large reflectors were not going to be the answer. The engines using these large reflectors were high temperature steam machines. It appeared that a better solution could be obtained if lower temperature media could be used making it possible to eliminate the high temperature, expensive, focusing reflectors and use low temperature solar collectors instead.

"Charles Tellier, a French engineer, was the first person in modern times to develop low-temperature solar collectors to drive machines. Whereas conventional engines used pressurized steam, Tellier's devices used pressurized vapor from certain liquids having boiling temperatures well below that of water. Ammonia hydrate, for example, will boil at −28 °F; sulphur dioxide will boil at 14 °F. These substances vaporize rapidly when exposed to higher temperatures.

Tellier began experimenting with these refrigerants as a means of powering a solar pump. On the sloping porch roof of his shop in Auteil, an exclusive suberb of Paris, he set up a row of ten solar collectors. They were metal plates, each four feet wide by eleven feet high, made of two sheets of corrugated iron riveted together. The grooves in the two sheets were aligned to form a series of hollow channels through which the ammonia hydrate followed. The bottom metal sheet of the collector was insulated to help block heat losses.

As the sun struck the top of the collectors, the metal conducted solar heat to the liquid inside. As a result, the ammonia vaporized and exerted a pressure of 40 pounds per square inch. The vapor circulated through pipes to a water pump consisting of a sperical chamber submerged in a well. The pressurized ammonia gas pushed a diaphragm in the chamber downwards, forcing water out of the pump in a jet. Afterwards the gas traveled through metal tubes set in a tank of cold water. The vapor condensed to a liquid again, and the ammonia was ready to repeat another cycle. According to Tellier, the pump drew over 300 gallons of water an hour."[11]

For reasons that are not certain, Tellier dropped his research on sun powered engines and returned to his work in refrigeration. However, his research and experimentation laid the ground work for the new concept of getting power from the sun.

"Two American engineers, H.E. Willsie and John Boyle, took up where Tellier had left off. Between 1892 and 1908 they explored the potential of low-temperature solar power plants based on the French inventor's design. In the May 13, 1909, issue of *Engineering News,* Willsie described what he and his colleague had accomplished. They had first begun to look into solar energy, he wrote, because they realized—as had Ericsson and Eneas before them—that the sun-drenched American Southwest desperately needed a source of cheap power for its irrigation pumps and mines.

Willsie and Boyle began their research with a review of the solar motors built by their predecessors. They discovered that 'the state of the art most developed included reflectors concentrating the sun's rays upon some sort of boiler.' But they also knew that every reflector-powered motor had been a commercial failure. Therefore, they chose to work with a Tellier-type motor using a low boiling-point liquid, so that a sophisticated and expensive reflector would be unnecessary."[12]

After ten years of experimental work and research, in 1903 they decided that they were ready to build a full scale sun power plant. The first was made in 1904 in St. Louis and incorporated a conventionally fueled heater to provide power at night or on sunless days. The success of this plant led them to transfer their company to Needles, California where the sun shines 85 out of 100 days. Here the last of several plants worked better than any solar power plant built up to that time. The cost of this plant was about $164 per horsepower compared to $40 to $90 for a conventional plant. Operating costs for the solar plant were considerably lower, particularly in the Southwest where the shipped in coal was prohibitively expensive.

This higher first cost and the introduction of the gas-producer engines that took place about this time were no doubt factors that led to the eventual ceasing of work by Willsie and Boyle. Available records do not indicate any further expansion and the experimentation ceased with the Needles plant. Willsie and Boyle had demonstrated that a sun-powered engine could run effectively using solar collectors instead of reflectors. They showed further that if a solar plant were combined with a conventional engine backup, the plant could run twenty-four hours a day.

In 1906 still another engineer entered the solar powered engine field. Frank Shuman of Philadelphia, Pennsylvania had established himself and had made a number of inventions and had an eye for technology of the future. He realized the finite limitation of fossil fuels and believed that solar energy was the only answer to supplying the world's energy needs. "One thing I feel sure of," he wrote, "and that is that the human race must finally utilize direct sun power or revert to barberism."[13]

After a successful demonstration run over two years of a solar powered pumping plant he had built in his back yard at his home in Tacony, a suburb of Philadelphia, Shuman persuaded a number of American investors to form the Sun Power Company, giving him sufficient funding to enable him to carry out some of his ideas for a large solar powered plant.

After considerable experimentation and research, resulting in several improvements in collector design and energy storage, he completed in Meadi, Egypt in 1913, a solar powered plant that produced 55 horsepower, sufficient to pump 6000 gallons per minute. The performance of this engine far surpassed that of any solar powered engine built up to that time.

Shuman spent seven years and about $250,000 on his work developing solar power. The successful plant at Meadi, Africa in 1913 encouraged him and seemed to confirm his earlier predictions that the large scale use of the sun to obtain power was about to become reality. "In February of 1914 he wrote,

> 'Sun power is now a fact and no longer in the "beautiful possibility" stage . . . It will have a history something like aerial navigation. Up to twelve years ago it was a mere possibility and no practical man took it seriously.
> The Wrights made an "actual record" flight and thereafter developments were more rapid. We have made an "actual record" in sun power, and we hope for quick developments.

Many others agreed, avidly supporting solar power. Some were former skeptics like those at Scientific American, who now praised Shuman's solar engine as 'thoroughly practical in every way.'

Besides the world science, Western Europe's colonial powers also lauded Shuman's work and looked forward to the enormous economic benefits of using solar energy in underdeveloped Africa. Lord Kitchener offered the Sun Power Company a 30,000-acre cotton plantation in the British Sudan on which to test solar-powered irrigation. The German government called a special session of the Reichstag to hear Shuman speak, an honor never before bestowed on an inventor. Speaking in the German he had learned as a boy, Shuman described the fantastic possibilities of solar power and showed movies of the Meadi plant at work. Duly impressed, the Germans offered $200,000 in Deutschmarks for a sun plant in German Southwest Africa. With such enthusiastic demonstrations of support, Shuman now expanded the scope of his plans. He hoped to build 20,250 square miles of reflectors in the Sahara, giving the world 'in perpetuity the 270 million horsepower per year required to equal all the fuel mined in 1909.'

But his grand dream disintegrated with the outbreak of World War I. The engineers running the Meadi plant left Africa to do war-related work in their respective homelands, as did Shuman who returned to the United States. He died before the war's end.

Gone was the driving force behind large-scale solar development. Moreover, with the Germans in defeat and their African colonies taken over by the Allies, the promises made to the Sun Power Company were as worthless as the Deutschmarks offered it. And the British, too, had lost interest in solar power. They began to turn towards a new form of energy to replace coal—oil. By 1919, the British had poured more than $20 million into the Anglo-Persian Oil Company. Soon afterwards new oil and gas strikes occurred in many parts of the world—southern California, Iraq, Venezuela, and Iran. These were almost all sunny areas where coal was difficult to obtain—areas targeted by Shuman, as well as Mouchot and Ericsson before him, as prime locations for solar power plants. But with oil and gas selling at near giveaway prices, scientists, government officials, and businessmen became complacent again over the energy situation, and the prospects for sun power quickly declined."[14]

The availability of cheap energy that came with the discovery of oil and natural gas marked the end of any further serious work on developing solar power. Until the oil embargo of 1973 with curtailment and reduction in the supply of oil and with the resulting rapid increase in the price of oil as dictated by the OPEC nations, there was little incentive for seeking alternative sources of energy. Now that has all changed.

## RECENT DEVELOPMENTS IN THE USE OF SOLAR ENERGY

The oil embargo awakened the Nation to the reality of what some scientists had been warning us about for sometime. Since 1947 the United States has been depending on outside sources for more and more of the oil necessary to sustain the ever increasing energy demands of the domestic or residential, the commercial and industrial sectors of our nation. The price of oil jumped from $3.00 to $12.00 per barrel during the 1973-74 embargo. In the ten year period from 1970 to 1980, the price went from $2.00 per barrel to $32.00 per barrel, a sixteen fold increase. Add to this the fact that with our present rate of oil consumption, the U.S. resources for conventional oil will be seriously depleted by the year 2000. No wonder that suddenly we have become an energy conscious nation. New governmental agencies have been formed; large sums of money have been allocated for research and implementation; and the nation is being mobilized to take steps to (1) find ways to conserve the energy we now have, (2) to develop known technologies to extend our present resources, and (3) to find and develop new alternative energy sources. One of these is solar energy.

As is evident from the historical sketch above, attempts to invent or devise methods and machines that would provide heat and power, supplanting oil and gas, were not new. The remainder of this paper will cover the development of the use of solar energy in the United States in three different sectors; the private or residential, the commercial, and the industrial. Some attention will be given to the earlier techniques developed to use solar energy in these three areas, the primary emphasis will be given to the efforts made to use the sun as an energy source since the crisis event of the oil embargo of 1973-74.

## THE USE OF SOLAR ENERGY IN THE RESIDENTIAL SECTOR

The use of solar energy to reduce the cost of heating man's dwelling places dates back thousands of years and many of the techniques used to accomplish this are those being employed today for the same purposes. Basically, it is a matter of positioning the home in such a way and incorporating architectural features to enable the rays of the sun to give maximum heat input to the building during the colder months and to reduce the amount of heat input during the warmer months. If this is done without the use of mechanical equipment, it is known as passive solar heating or cooling. When collectors incorporating circulating pumps, heat exchangers, etc. are used, it is referred to as active solar heating or cooling. In the very early Greek and Roman times, referred to earlier, the placement of houses and buildings to take maximum advantage of the sun's energy made use of what is now termed passive design. Although concentration of the sun's rays by mirrors or reflectors had been discovered, as far as can be determined little or no practical use was made of this more concentrated, higher temperature heat source.

Through the years there have been many instances where the sun's energy has been harnessed to aid in the heating of buildings and in the heating of water for domestic purposes. The advent of low cost fossil fuels, namely oil and natural gas, all but eliminated the further development of practical solar energy. During the early decades of the 1900's, from 1910 to 1940, there were numerous individuals, engineers, and architects who designed and built homes, schools, and apartments where attempts were made to capture the sun's energy and reduce the need for the fossil fuels, coal, oil and gas, to heat these buildings. These attempts were successful in varying degrees, depending on the type of building, its orientation, and provisions made to take advantage of this inexpensive form of energy. By the late 40's the abundance of oil and gas all but eliminated further attempts to harness the direct energy from the sun to supply man's need for energy.

Renewed efforts to again seek ways to take advantage of the sun's energy potential came as the result of the oil embargo in 1973-74, and in the realization that at current consumption rates, there was enough oil and gas for only a few more decades.

New agencies were established by the government and funds were allocated to encourage the development of ways and means to use the direct radiation from the sun to provide at least a portion of the energy needed for man's necessities.

In addition to funding research and development, the government is trying to encourage the constituency to apply the knowledge and technology developed and purchase and install the equipment necessary to make use of this available energy. Tax incentives and special low cost loans are available to the home owner who will reduce his use of the fossil fuels by using solar energy. Incentives for home owners to install solar energy systems have come from both state and national sources. "Since 1974, forty-three states have passed some form of tax incentive for solar energy systems. And, recently, the U.S. Congress enacted major solar energy incentive legislation as a part of the National Energy Act."[15]

## THE CURRENT USE OF SOLAR ENERGY IN THE RESIDENTIAL, COMMERCIAL AND INDUSTRIAL SECTORS

It is difficult, in fact, it may well be next to impossible, to obtain accurate data on the extent to which solar energy is being used in the three sectors where it is most applied; namely, the residential, commercial, and industrial sectors. Installations are frequently made by independent contractors and installers and, particularly in the residential sector, by "do-it-yourself" installers, and often very limited or no records are kept of the installations.

Some idea of the recent growth in the use of solar energy for residential use can be gained from the Solar Fact Sheet, 7th Edition, No. 106, for February 1980, published by the National Solar Information Center.[16] This fact sheet, using a sampling of available data from various sources, gives chronologically the progress of solar application in homes and buildings from July 1974 through March 1979.

| DATE | NO. OF BUILDINGS | SOURCE |
|---|---|---|
| July 1974 | 52 solar heated | W.A. Shurcliff, *Solar Heated Buildings: A Brief Survey,* 4th Edition |
| May 1975 | 118 solar-heated buildings | Shurcliff, 9th Edition |
| Dec. 1975 | 183 solar-heated homes | *Sunup,* February 1978 |
| Sept. 1976 | 1500 solar heated buildings by end of 1976 | UPI, quoting "government experts |
| Early 1977 | 300-400 solar-heated/ cooled buildings; 5000-10,000 solar hot water heaters | SEIA cited Copper Development Association, Inc., fact sheet |
| January-June 1977 | 12,000-24,000 new solar houses built or retrofitted | *Solar Energy News,* May 4, 1978 |
| Oct. 1977 | 6000 solar homes | Fred Dubin, in a speech for the Mid-Atlantic Solar Energy Association |

| | | |
|---|---|---|
| End of 1977 | 40,000 residential installations of solar systems were made in 1977 | Frost and Sullivan promotional flyer |
| Feb. 1978 | Over 5000 homes have solar heat | *Sunup,* February 1978 |
| Nov. 1978 | California has 3000 solar heated and/or cooled homes and buildings. 11,000 solar domestic hot water systems. | *Toward a Solar California,* Solarcal Council of California, January 1979 |
| End of 1978 | Unofficial Census reports 1% of all new housing starts in 1978 had solar installation. (Statistical methods used to obtain this figure have been criticized) | *Solar Engineering,* Sept. 1979 |
| March 1979 | Over 50,000 solar dwellings already exist. | *Saturday Review,* March 3, 1979 |

The several different sources differ in specific numbers but if taken as a whole, there is a definite trend upward in the number of applications of solar energy use in residential homes.

The most comprehensive estimate of existing solar activity is found in the Solar Energy Institute of North America's State of the Union Report of 1979.[17] The figures for 1978 and projections for 1979 were compiled from solar organizations, state energy offices, consultants and solar professionals. Their totals for the nation as a whole are as follows:

TOTALS

| | | |
|---|---|---|
| Systems installed in 1978 | 33,000 | |
| Existing at end of 1978 | | 88,000 |
| Projected installations for 1978 | 50,000 | |
| Projected by end of 1979 | | 140,000 |

1978 INSTALLATIONS BY SYSTEM TYPE

| | |
|---|---|
| Domestic Hot Water Only | 27,000 |
| Space Heat and Hot Water | 3,500 |
| Space Heat Only | 2,000 |
| Cooling | 500 |

Since most solar collectors are made by manufacturers, an indication of the extent to which solar energy is used can be obtained by knowing the area of the collectors manufactured. This would apply to commercial and industrial applications as well as the residential. The uses for solar energy in the commercial sector are essentially the same as for the residential, primarily to heat hot water and

TABLE 1

*Solar Collector Manufacturing Activity*

| | Area (Thousands of Square Feet) | | |
|---|---|---|---|
| Year | Low Temperature | Medium Temperature | Total |
| 1974 | 1137 | 137 | 1274 |
| 1975 | 3026 | 717 | 3743 |
| 1976 | 3876 | 1925 | 5801 |
| 1977 | 4743 | 5569 | 10312 |
| 1978 | 5872 | 4988 | 10860 |
| 1979 | 8395 | 5857 | 14252 |
| 1980 | 11928 | 7527 | 19455 |

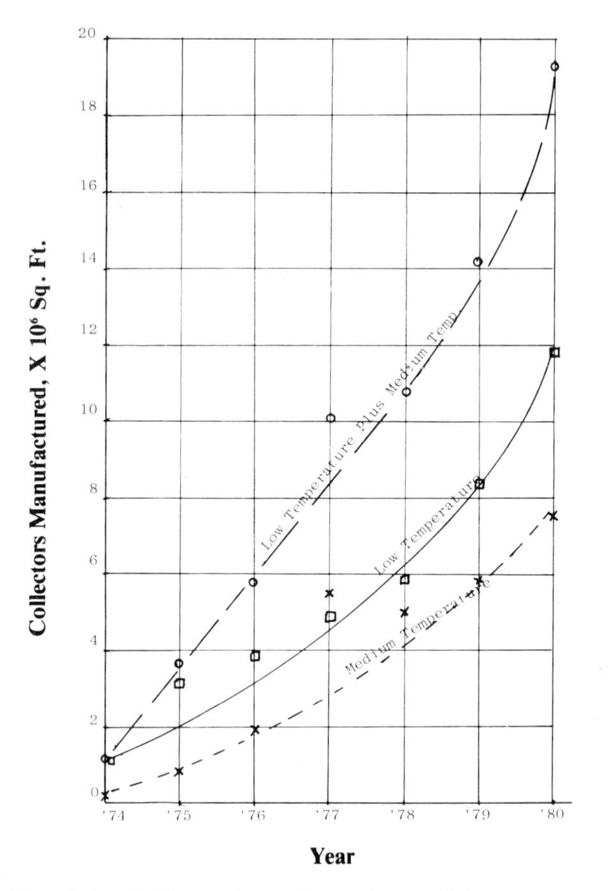


FIGURE 1. Solar Collector Manufacturing Activity

for space heating. In most instances the amount of heat needed is a function of building size and use. Commercial buildings including schools, office buildings, churches, hospitals, etc., would have larger heat requirements which would be reflected in the size of the solar installation to provide this heat.

The summary Table 1 gives the thousands of square feet of collectors manufactured in the United States from 1974 through 1980.[18] The information is abstracted from a more inclusive table in the reference cited. The collectors are classified as low temperature and medium temperature. Low temperature collectors generally operate at a temperature of 110 °F or less. It has no glazing and is generally made of plastic or rubber. The medium temperature collector operates in the 140 to 180 degree F. range but may go as low as 110 °F. It would have one or two glazings, a metal frame, a metal absorption panel, and insulation on the sides and back of the panel.

In 1980, the low temperature collectors accounted for 61% of all collector shipments and 94% of these were used for heating swimming pools. Figure 1 is a plot of the information in Table 1. It is interesting to note that the yearly increase is linear through 1978. Significantly larger incerases in 1979 and 1980 are shown resulting from greater numbers of low temperature collectors manufactured.

Summary Table II gives the breakdown of the collector areas used in the residential, commercial and industrial sectors for the three years 1978, '79 and '80.[19] As in the previous table, the information given here is abstracted from a more inclusive table shown in the reference cited. The residential sector is using solar heating at an increasing rate that is greater than that for the commercial and industrial sectors. This original breakdown of these figures indicates that in 1980 twice the number of low temperature collectors were sold as compared to the medium temperature collectors. This is accounted for by the very common use of low temperature collectors for heating swimming pools.

These figures lead to some interesting observations. In 1978, 77% of the collectors made went to the residential sector, 17.6% to the commercial sector, and 1.6% to the industrial sector. In 1979, 79.9% went to residential, 14.1% to commercial and 2.2% to industrial users. In 1980, 82.8% went to residential, 12.5% to commercial, and 2.63% to industrial uses. In each of the three years, the residential sector used a greater portion of the collectors made, increasing from 77% to 79.9 to 82.8%. During that same period the commercial sector decreased its fraction of the total from 17.6% in '78, 14.1% in '79, and 12.5% in 1980. The industrial sector increased slightly going from 1.6% in '78, to 2.2% in '79, and to 2.63% in 1980.

The reasons for this are not entirely clear but it may reflect the additional efforts to promote the residential use of solar heat. It may also mean that the potential market for solar is greater in the residential sector. Although the number of collectors used in industry is relatively very small, the fraction of collectors going to industry increased with each succeeding year. In terms of

TABLE II

*Solar Collector Applications According to Market Sectors (Thousands of Square Feet)*

| Year<br>Collector Type | Residen-<br>tial | Commer-<br>cial | Indus-<br>trial | Agricul-<br>tural | Other | Total |
|---|---|---|---|---|---|---|
| 1980 | | | | | | |
| Low-Temp, Non-Metallic | 8168 | 1034 | 112 | 35 | 0 | 9349 |
| Low-Temp, Metallic | 2030 | 457 | 33 | 16 | 24 | 2560 |
| Medium Temp, Air | 555 | 43 | 1 | 15 | 7 | 620 |
| Medium Temp, Lquid | 5216 | 710 | 274 | 19 | 203 | 6422 |
| 1979 | | | | | | |
| Low-Temp, Non-Metallic | 6064 | 756 | 115 | 11 | 0 | 6946 |
| Low-Temp, Metallic | 1057 | 170 | 22 | 5 | 193 | 1447 |
| Medium Temp, Air | 735 | 80 | 3 | 21 | 7 | 846 |
| Medium Temp, Liquid | 3377 | 764 | 93 | 31 | 195 | 4460 |
| 1978 | | | | | | |
| Low-Temp, Non-Mctallic | 4198 | 577 | 23 | 60 | — | 4858 |
| Low-Temp, Metallic | 740 | 124 | 2 | 1 | 24 | 891 |
| Medium-Temp, Air | 538 | 91 | 10 | 35 | 50 | 724 |
| Medium-Temp, Liquid | 2546 | 858 | 167 | 17 | 59 | 3647 |

numerical figures, all three sectors showed marked increases in the area of collectors used each year.

## THE FUTURE OF SOLAR ENERGY IN THE UNITED STATES

Some idea of the potential uses of solar energy by the United States can be gained from the following quotations. President Carter on June 20, 1979, when announcing the solar policy and dedication of the White House Solar System said:

> "By the end of this century, I want our Nation to derive 20 percent of all the energy we use from the sun—direct solar energy in radiation and also renewable forms of energy derived more indirectly from the sun. This is a bold proposal, and it is an ambitious goal. But it is attainable if we have the will to achieve it."[20]

Predictions have been made by various individuals and organizations as to the amount of energy the United States will derive from solar energy by the end of the century.

> "ERDA" believes that solar energy technology offers the potential for supplying as much as 25% of the nation's future energy needs from domestic resources by 2020 if costs of collecting and utilizing solar energy can be reduced substantially."

A number of findings resulted from the Solar Energy Research Institute's Solar Conservation Study recently completed.

> "The energy which can be saved in buildings in the United States represents the largest and least expensive source of energy that can be supplied during the next two decades. Buildings now consume about a third of all energy used in the United States using 13 Quads per year (Quadrillion Btu) of oil and gas and 13 additional Quads per year of primary energy to generate electricity."[23] (used in these buildings)

Of the approximately 26 quads of energy going into residential and commercial buildings, the Solar Energy Research Institute's study projects that the application of solar energy technology to new buildings could save from 0.3 to 0.5 quads. Retrofitting older homes using a wide range of approaches and technologies to incorporate solar heating is estimated to displace 0.8 quads by the year 2000.

The potential for using solar energy for domestic and commercial hot water is even grater than that for space heating. The technology is readily available and as the cost of electricity, oil, and gas continue to rise, it becomes increasingly desirable to use solar energy to supply most of the heat for hot water. The Solar Energy Research Institute estimates that this could displace 0.5 to 0.6 quads by the turn of the century.

If those areas where small wind machines can be successfully used would develop this technology, it is estimated that 0.8 to 1.1 quads could be displaced by 2000.

It is predicted that photovoltaics will, by the year 2000, be extensively used in new homes with the potential of displacing 0.3 to 0.45 quad of conventional fuel energy.

Based on these predictions by the Solar Energy Research Institute and adding the various contributions, it is possible that solar energy could displace from 2.7 to 3.45 quads. Of the 26 quads going into residential and commercial buildings, this would be a savings of 10.4 to 13.3 percent.

The Mitre Corporation in 1978 reported on a study entitled "A Comparative Analysis to the Year 2020" saying,

> "By 1985 the projected installation of 1.7 million hot water and space heating systems should result in a solar contribution of 0.15 quads of energy per year . . . (by 2000) over 16 million hot water, heating and cooling systems are expected to result in a solar contribution of 1.6 quads . . . By the year 2020 solar energy might displace the equivalent of nearly 18 percent of the nation's 189 Quads fuel—33 to 35 quads from solar. This is equivalent to about 45 percent of the fuel consumed today."[24]

In the report of the energy project at the Harvard Business School, entitled "Energy Future" edited by Stobaugh and Yergin, the chapter "Solar America" comments on the contribution that solar energy will make toward future energy needs:

> "How much of America's energy needs could be derived from solar energy by the year 2000? Estimates vary over a very wide range and comparative analysis is difficult, because different analysts work with different definitions of solar. Some begin and end with solar heating, whereas others include all the solar options except hydropower. Recently, however, many analysts have begun to converge on a definition similar to the one presented at the beginning of this chapter, which includes all the "recent" solar cycles. Even after adjusting predictions to our definition of solar energy, a selection of high-quality projections still ranges from 7 to 23 percent by the year 2000.
>
> The truth is that no one really knows what the contribution from solar energy will be in the year 2000. Solar's contribution depends on at least six other complex and uncertain variables, each difficult to forecast.
>
> 1. Prices of competing energy sources
> 2. Overall levels of domestic energy consumption
> 3. Level of federal involvement in solar energy
> 4. Rate of advancement of solar technologies
> 5. Rate at which institutional barriers to solar will be overcome
> 6. Reliability of energy supply
>
> In this context of uncertainty, not uncommon for new technologies, predictions are fraught with risk. To have a high probability of being correct one can only speak in terms of

ranges. For instance, it is relatively safe to predict that solar energy by the year 2000 will be 10-30 percent of the nation's energy supply.

On the other hand, it is difficult to motivate people toward a range of goals. For this reason, managements and governments set specific goals. To be effective, goals must be difficult to reach but achievable. It is in this spirit that I believe that a goal of achieving one-fifth of the nation's energy from solar by the year 2000 is a sensible goal for our country. This does not mean that uneconomic solar technologies should be espoused, but it does mean that the portfolio of solar technologies includes alternatives which, if given a fair chance against conventional sources and an initial stimulus, may contribute up to 10 mbd of oil equivalent by the turn of the century.

Presently solar energy in the form of wood and hydropower contributes 6 percent of the national energy supply. A tripling in solar's contribution over a twenty-year period is not without precedent in the nation's energy history. Earlier in the century, the use of gas and oil tripled and quadrupled, respectively, over a twenty-year period.

But to achieve a 20-percent contribution from solar by the end of the century, a commitment must be made now to existing technologies. The only realistic two options for the short-term are wood and wood waste, and on-site solar technologies, such as solar heating, small hydro-power, and small wind. The short-term challenge is not technology, but accelerating diffusion. As stated before, we recommend self-extinguishing tax incentives to get these technologies through the early sluggish phase of commercialization, combined with measures to overcome the institutional barriers."[25]

Although it is not possible to predict precisely the contribution that solar energy will make toward satisfying the energy needs in the residential, commercial and industrial sectors, there is overwhelming evidence that the use of direct solar energy is a viable, plausible, and effective source of energy that becomes more and more attractive as the cost and availability constraints of gas and oil make themselves felt. Continued application of present technology to make available solar energy and continued research and the development of new and improved means for more effectively utilizing this unlimited source of energy will contribute significantly to the conservation of other forms of energy and making the United States less dependent on foreign sources of energy.

## NOTES

1. William W. Eaton, "Solar Energy" in *Perspectives on Energy* by Lon C. Ruedisili and Morris W. Firebough (New York, Oxford University Press, 1978), p. 418
2. Ken Butti and John Perlin, *A Golden Thread* (Palo Alto, California: Cheshire Books, 1980), p. 3
3. Ibid., p. 13
4. Ibid., p. 15
5. Ibid., p. 27
6. Ibid., p. 29
7. Ibid., p. 34
8. Ibid., p. 63
9. Ibid., p. 67
10. Ibid., p. 74
11. Ibid., p. 94
12. Ibid., p. 95
13. Ibid., p. 101
14. Ibid., p. 109-111
15. U.S. Department of Housing and Urban Development, *Solar Status* (Washington, DC, 1980, 623-278/1366), p. 1
16. National Solar Information Center, *Solar Fact Sheet* (F.S. 106, 7th Edition, February 1980), p. 1
17. Solar Energy Institute of North America, *Solar State of the Union Report 1979,* Washington, DC, 1979.
18. U.S. Department of Energy, *Solar Collector Manufacturing Activity, July through December 1980,* (DOE/EIA - 0174 (8012), March 1981), p 3
19. Ibid
20. National Solar Information Center, *Solar Fact Sheet* (F.S. 106, 7th Edition, February 1980), p. 3
21. Ibid
22. Ibid
23. Solar Energy Research Institute (SERI) Solar Conservation Study, 1981, *A New Prosperity, Building a Sustainable Energy Future,* Andover, Massachusetts, Brick House Publishing, 1981, p. 11
24. Mitre Corporation, *Solar Energy: A Comparative Analysis to the Year 2000,* 1978. Report HCP/T 2322-01, National Technical Information Service.
25. Robert Stobaugh and Daniel Yergin, *Energy Future: Report of the Energy Project at the Harvard Business School,* New York, Ballantine Books, pp. 262-265.

## REFERENCES

Butti, Ken, and Perlin, John. 1980 *A Golden Thread.* Palo Alto, California: Cheshire Books

Kreider, Jan F., and Kreith, Frank. 1981. *Solar Energy Handbook, New York, NY, McGraw Hill.*

*Mitre Corporation, 1978, Solar Energy: A Comparative Analysis To the Year 2000.* Report HCP/T 2322-01, National Technical Information Services.

Montgomery, Richard H. and Budnick, Jim, 1978. *The Solar Decision Book,* New York, NY, John Wiley and Sons.

National Solar Information Center, February 1980, *Solar Fact Sheet.* (F.S. 106, 7th Edition)

Ruedisili, Lon, C., and Firebaugh, Morris W., 1978. *Perspective On Energy.* New York: Oxford University Press

Solar Energy Institute of North America (SEINAM) 1979, *Solar State of the Union Report for 1979,* Washington, DC

Solar Energy Research Institute, 1981, *Solar Conservation Study* "A New Prosperity, Building a Sustainable Energy Future" Andover, Massachusetts, Brick House Publishing.

Stobaugh, Robert, and Yergin, Daniel, 1979. *Energy Future: Report of the Energy Project at the Harvard Business School,* New York, NY Ballantine Books.

Stoker, H. Stephen, Scager, Spencer L. and Capener, Robert L., 1975. *Energy from Source to Use,* Glenview, Illinois: Scott, Foresman and Company.

U.S. Department of Housing and Urban Development, 1980. *Solar Status.* 623-278/1366. Washington, DC

*Chapter Nineteen*

# A Summary of the Radiological and Physical Conditions of the Three Mile Island Unit-2 Reactor Containment and Their Impact on Recovery

**James E. Tarpinian**
BECHTEL NORTHERN CORP.
P.O. Box 72, Middletown, PA 17057
**Peter Hollenbeck**
GENERAL PUBLIC UTILITIES
P.O. Box 480, Middletown, PA 17057

James Tarpinian holds a B.A. in Biology from the University of Connecticut at Storrs and an M.S. in Radiological Health Physics from the University of Lowell in Lowell, Massachusetts. His health physics career began at Electric Boat Division of General Dynamics. Mr. Tarpinian is currently a Site Liaison Engineer with Bechtel Northern Corporation, the prime contractor for the Three Mile Island recovery, and is involved with the Data Acquisition Program and reactor building decontamination. Mr. Tarpinian is a member of the Delaware Valley and national chapters of the Health Physics Society.

Peter Hollenbeck holds a B.S. in Radiological Health Physics from the University of Lowell in Lowell, Massachusetts. His experience in health physics was gained at New England Nuclear Corporation and Maine Yankee Atomic Power Company. He is currently a radiological engineer with General Public Utilities' Radiological Technical Support Group, and performs the radiological reviews and evaluations of all reactor containment entry work. Mr. Hollenbeck is a member of the national chapter of the Health Physics Society.

## INTRODUCTION

The first entry into the reactor containment building at TMI-2 occurred on July 23, 1980, sixteen months after the accident. This event marked a milestone in the recovery of the damaged reactor and containment building. Until that point, information about the radiological and physical conditions of the building was sparse as measurements had to be obtained remotely. Since then, entries have been made regularly in order to obtain data crucial to the planning and engineering of the recovery effort.

The recovery program is divided into three phases. Phase I is devoted to the design and construction of the facilities to support recovery, the processing of the 600,000 gallons of contaminated water in the containment basement, and the decontamination of the reactor building surfaces.

Phase II milestones will include the removal of the reactor vessel head, the removal of the fuel and reactor vessel internal structures, and the decontamination of the reactor coolant system.

The objective of Phase III is to complete the work necessary to requalify and license the plant for operation. The culmination of Phase III is plant start-up. At this time General Public Utilities (GPU) has not committed to the restart of TMI-2. The primary concern is for the safe removal and storage of the fuel in the reactor vessel, and the present efforts are directed toward this goal.

In addition to the planning and engineering associated with each phase, current efforts include the Data Acquisition Program as a portion of Phase I. The purpuse of the Data Acquisition Program is to obtain quantitative and qualitative information about the radiological and physical characteristics of the building. The objective of this paper is to review the gathered information to date and to examine the impact the conditions of the building will have on the recovery. Appendix A contains a description of the radiological terms used in this discussion.

## DESCRIPTION OF THE REACTOR BUILDING

Prior to the discussion it will be helpful to describe the layout of the building. A cross-section of the building is shown in Figure 1.

The building is constructed of reinforced concrete 48 inches thick, and is lined inside with a carbon steel plate. The reinforced concrete floors at each level are nominally 8″ thick, but may be up to 36″ thick in some places. The floors, walls, and other surfaces are painted with a coating specifically selected to withstand damage from the insult of heat, pressure, and radiation resulting from the design basis loss-of-coolant accident (LOCA).

The reactor building basement floor is located at elevation 282′6″ above sea level. During the accident steam and water were discharged to the building through the rupture disc on the reactor coolant drain tank located on this elevation. The steam and water were contaminated with radioactive fission products

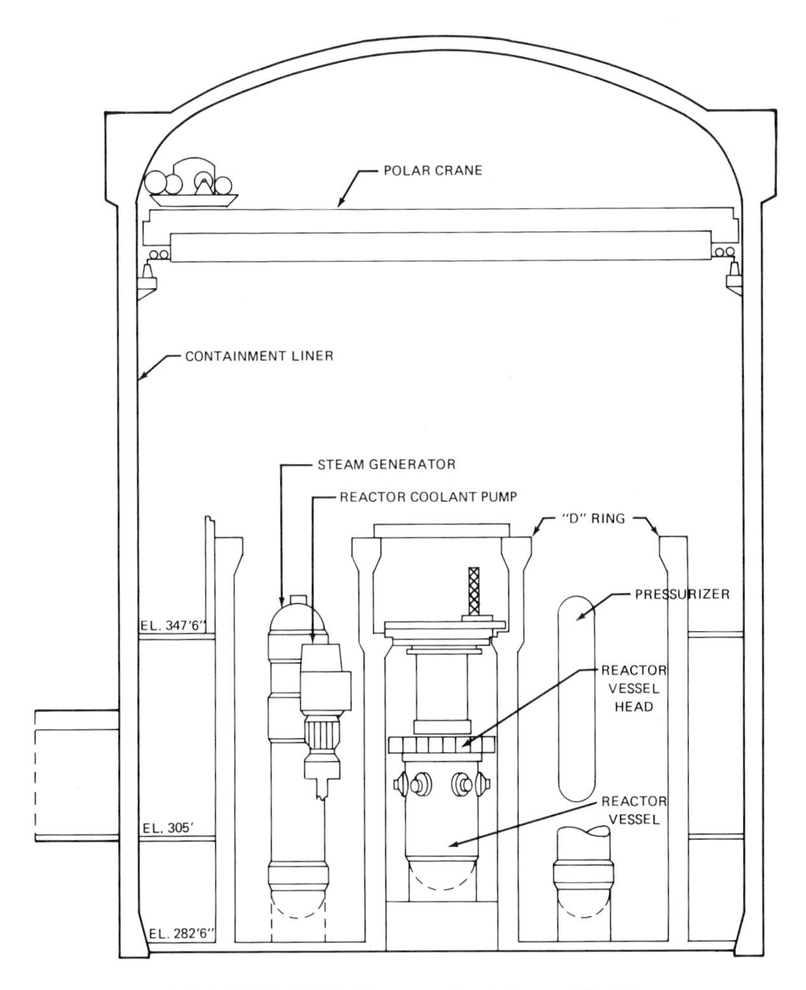


FIGURE 1. TMI-2 Reactor Containment Building

which were released to the coolant as a result of the failure of the fuel cladding. The steam provided a transport mechanism for the contamination to be dispersed throughout the building, where it subsequently became deposited on the reactor building surfaces. The water still resides in the basement and is about eight feet deep.

The 305′ elevation is the entry level of the building. Currently, personnel enter and exit through the personnel airlock #2 shown on Figure 2.

El. 347′6″ is the focus of Phase II defueling activities. Access to the 347′6″ elevation can be made by use of either stairwell. The polar crane, shown on Figure 1, will be used to lift the reactor vessel head and place it on its storage stand. Removal of the head allows access to the fuel assemblies.

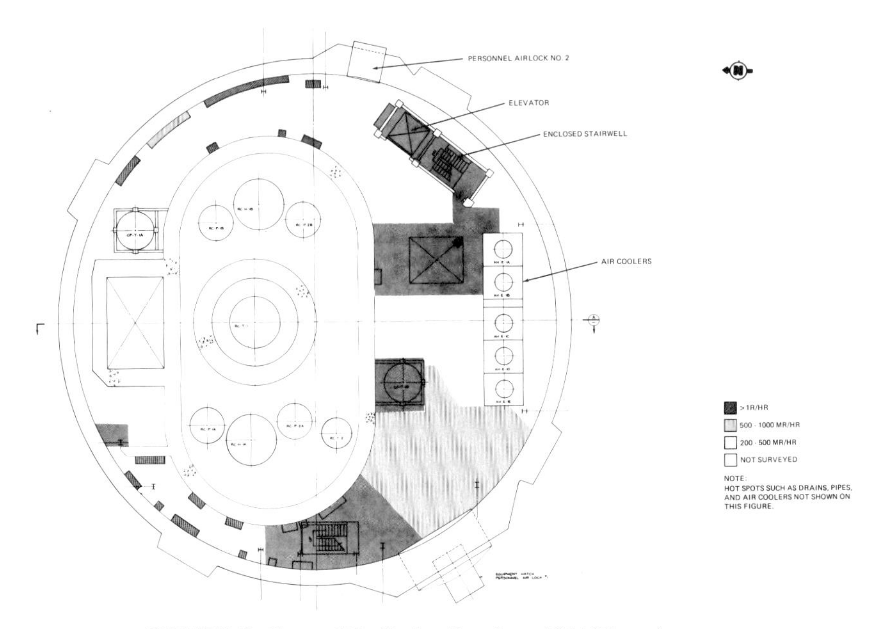


FIGURE 2. General Radiation Levels at 305′ Elevation

## PHYSICAL CONDITIONS

The building does not appear to have suffered significant structural damage. The most extensive damage is the rusting of exposed steel surfaces due to prolonged exposure to water and moisture. Some minor damage occurred as a result of the pressure transient during the accident. In addition, there is localized thermal or chemical damage and some minor damage possibly due to radiation effects on materials.

The floors of the building have a layer of dirt and grit and there are frequent patches of rust stains. Water and moisture have caused considerable corrosion of exposed steel surfaces, however most structural steel and sheet metal are painted. Some flaking of the paint has occurred on the steel liner above El. 347′6″. The paint on the "D" ring walls, concrete block, and most equipment is in good condition. Water marks show evidence of puddling on the floors. Water streaks are evident on the "D" ring walls and containment liner.

Stainless steel surfaces in the building include the refueling pool liner, the reflective insulation and some piping. These surfaces are not discolored and appear structurally normal.

Much of the equipment in the building appears to be in fair condition although it is expected that unsealed electrical motors may be moisture damaged. The polar crane is not operable although the extent of the damage has not been fully

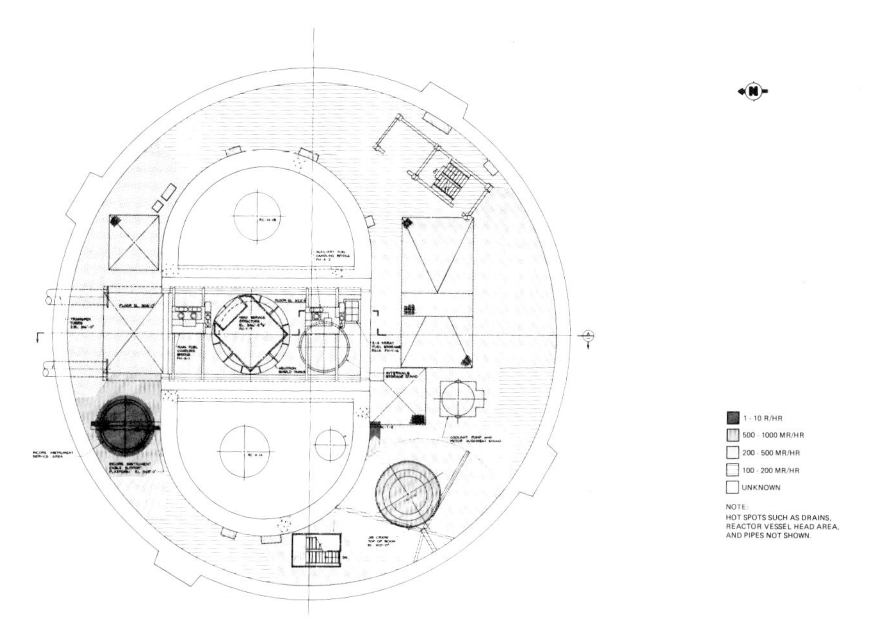


FIGURE 3. General Radiation Levels at 347.6′ Elevation

evaluated. One of the three conductors which transmit power to the crane has broken away from its rubber grommets and fallen to the floor on El. 347′. The elevator, which is one foot above its normal position on El. 305′, does not operate.

The upper elevator doors on El. 305′ and El. 347′6″ are bowed several inches toward the building interior as a result of the pressure transient. Additional related damage includes the door to the enclosed stairwell on El. 305′ which was forced open beyond its normal resting place. Several 55-gallon steel drums on El. 347′6″ were crushed or imploded as well. The pressure reached a maximum of 28 psig.

A number of items on both elevations suffered apparent thermal damage or possible chemical damage resulting from the sodium hydroxide in the building sprays which were activated during the accident. These effects are very localized and occur most frequently in various locations on El. 347′6″. These items are predominantly plastic and include a telephone, some rope, portions of small diameter electrical cord, tubing, and some pieces of wood. Several of these items have been removed for analysis.

Radiation has caused discoloration of gage glass, plastic electrical tie wraps, and fluorescent light covers. Some plastic items and small diameter electrical cord have become embrittled due to thermal or radiation effects. The total radiation dose to several of these items is being determined.

The lighting level in the building is good. Practically all of the lights are operational and no relamping has been necessary.

Most electrical power outlets on El. 347′6″ are operational, but all outlets in the building having electrical panels on El. 282′6″ are assumed to be non-operational. These include outlets on El. 305′. Electrical power is available on El. 305′ at the closed circuit television cameras discussed in a later section.

## RADIOLOGICAL CONDITIONS

The major sources of radioactivity contributing to the radiation levels in the building are:

1) the water on El. 282′6″
2) the contamination deposited on building surfaces, and,
3) the reactor coolant system.

Radiation measurements have been taken using portable survey instruments during each entry into the building, and the radiation levels have been fairly well established for most accessible areas. Figures 2 and 3 illustrate the general area gamma exposure rates on El. 305′ and El. 347′6″, respectively. It should be noted that the boundaries of the cross-hatched areas are not as well defined as shown since the exposure rates vary considerably within a few feet and the resulting gradient makes it difficult to draw a precise isodose map.

The dose equivalent rates on El. 305′ are generally 200 mRem/hr to 700 mRem/hr due to gamma radiation, and 200 mRem/hr to 1000 mRem/hr due to beta radiation. The presence of the water in the sump greatly influences the gamma radiation levels at this elevation. This is demonstrated by the higher radiation levels near the stairwells and floor penetrations shown in Figure 2. The floor provides some shielding from the gamma radiation emanating from the water.

The air coolers have accumulated contamination, due to the intake and recirculation of air, causing gamma radiation levels up to 2 Rem/hr. The floor drains have also accumulated contamination causing local beta radiation levels up to 54 Rem/hr. Aluminum shield covers have been placed over the floor drains to reduce the beta radiation levels.

The radiation levels on El. 347′6″ are somewhat lower than those on El. 305′ overall. Gamma dose equivalent rates are generally from 100 mRem/hr to 500 mRem/hr while beta dose equivalent rates are about 300 mRem/hr to 800 mRem/hr. The additional distance from the basement water and shielding of the floor helps to reduce the gamma radiation levels, but the contamination on the floor and liner is a significant contributor.

Samples of the contamination have been obtained on El. 305′ and El. 347′6″. The results of the analyses have shown the most abundant isotope to be Cs-137, with the isotopes Cs-134 and Sr/Y-90 present in smaller concentrations. Cs-137 and 134 are mainly responsible for the gamma radiation levels in the building, while all of the above isotopes contribute to the beta radiation levels.

Loose surface contamination as measured using the accepted practice of paper smears ranges from $10^{-3}$ to $10^{-1}$ uCi/cm$^2$ on the floors due to beta and gamma emitters. Loose surface contamination concentrations on the liner and "D" ring walls are about a factor of 100 smaller.

The ratio of Cs-137 to Cs-134 on the smear samples were generally between 6 and 7. The ratio of Cs-137 to Sr-90 exhibits greater variability, ranging from 5 to 200.

Gross alpha analyses of smears and other samples have shown only trace amounts of alpha emitting isotopes associated with fuel.

Analyses of samples from the water in the basement have shown concentrations of about 143uCi/ml of Cs-137, 19uCi/ml of Cs-134, and 5uCi/ml of Sr-90.

## RADIOLOGICAL PROTECTION OF PERSONNEL

Personnel entering the reactor building must be adequately protected from inhaling the particulate contaminants and monitored to assure that they do not exceed the administrative whole body quarterly limit of 1 Rem. Each person also wears protective clothing sufficient to prevent contamination from contacting the skin.

The protective clothing includes two (2) pair of cotton coveralls with hoods. This completely covers the body with the exception of the hands, feet, and face. The face is protected by the respiratory equipment described below. The feet are protected by rubber boots and multiple layers of plastic boot covers while the hands are protected by cotten gloves, several pair of surgical gloves, and at least one pair of rubber gloves. In addition, the individual may be required to wear a whole or partial set of plastic rain gear.

The clothing not only protects against skin contamination but also serves to shield against all but the most energetic beta radiation. Normally beta radiation does not contribute to the whole body radiation dose equivalent since it is usually absorbed in the dead layer of skin. In the reactor building skin doses due to beta have not been limiting, and while the beta dose rates are high, skin doses have been very low.

The respiratory protection for personnel is usually supplied by a battery powered air purification unit. The batteries, motor, and high efficiency filters are contained in a pack which mounts on a belt. The air is supplied to a full face mask connected by a hose. These units will reduce the air-borne concentration of respirable particles by a factor of 1000.

Each individual wears dose monitoring devices in several locations on the body. One such device has an LED display and measures accumulated whole body dose. This enables the person to conveniently monitor his dose while in the building since the display can be easily read through the mask's facepiece and under conditions of low light.

Communications are maintained via two way radio and closed circuit television (CCTV), and the entry is controlled from a command center. The CCTV system was installed in February, 1981 and includes four (4) cameras each on El. 305′ and El. 347′6″. This has proven invaluable for monitoring building entry activities and especially for planning and training for each entry.

## IMPACT ON RECOVERY

Currently the major efforts are oriented toward the examination of the fuel in the reactor core prior to removing the reactor vessel head. The data acquisition program has provided valuable information that will be used to guide the planning and engineering of these efforts. The radiological and physical conditions of the building are not as severe as suspected following the accident. Personnel can safely enter the building and engage in activities for prolonged periods of time.

The radiation levels in the building will be reduced by the processing of the water at El. 282′6″ and the decontamination of the building surfaces. The water will be pumped out of the building using a submersible pump that was installed in late April, 1981. Processing of the water will be accomplished using a series of ion-exchange resin filters, and the processed water will be stored on-site. Some of the processed water may be used for the decontamination. Removal of the water and surface contamination will enable personnel to work in the building for longer periods of time and will also help to minimize the radiation exposure incurred by the workers. The defueling activities cannot be performed without some decontamination of El. 347′6″.

One goal of the decontamination will be to enable the refurbishment of the polar crane. Fall prevention safety equipment was installed on the crane in May, 1981. A subsequent inspection and photographic survey was performed in July. The crane must be placed into service in order to perform the reactor vessel head lift necessary to gain access to the reactor core.

The TMI-2 recovery program has and will continue to provide a valuable source of information which will ultimately benefit the nuclear industry in the design, construction, operation, and decommissioning of commercial nuclear power reactors.

## APPENDIX

### Description of Radiological Terms

The terms used to describe radioactivity and radiation exposure are often confusing to those not directly involved with the use of radiation or radioactive materials. For the purpose of clarification it will be useful to some readers to have some information regarding the quantities and units used in this paper.

First of all it is important to distinguish between radiation and radioactivity.

"Radioactivity", or radioactive decay, is the process by which an energetically unstable atom seeks a more stable state. It rids itself of this excitation energy by emitting "radiation". Thus radiation is the product of radioactive decay.

There are many forms of radiation, but the types of radiation pertinent to this discussion are alpha, beta, and gamma. Alpha and beta radiation are particles with mass. An alpha particle is relatively heavy and has a positive charge. Because of this, even the most energetic alpha cannot penetrate an ordinary piece of cardboard. Beta particles are about 8,000 times less massive than alpha particles and have a negative charge. Like alpha particles, most beta particles cannot penetrate the dead layer of skin. The most energetic beta particles may travel up to thirty or forty feet in air. But even these may be stopped by a thick piece of rubber or a thin sheet of aluminum.

Gamma radiation is nothing more than high energy light particles, or photons. Photons have no charge, no mass, and travel at the speed of light. These physical properties cause gamma radiation to be very penetrating, that is, a single gamma photon may pass entirely through the body without interacting or causing any damage.

Radiation interacts in the body in such a manner as to transform the kinetic energy of the particle into chemical energy. This chemical energy in turn causes some amount of biological damage. This damage may be great or small depending upon the amount of radiation causing it. For this reason a system has been devised to measure radiation, and the units of expression ultimately relate to a relative amount of biological damage.

The quantity measured is called the "dose equivalent". The basic unit of this system is the "rem". A single dose of 1,000 rem would almost certainly kill a human being. We are all exposed to a certain amount of radiation due to natural radioactivity originating from the earth, outer space, and the water and food we consume. This background radiation amounts to an average of about 100 milli rems per year per individual. (One millirem equals 1/1000 of a rem). The limit of exposure to radiation workers in the U.S. has been set at 3 rems per quarter. The administrative guideline for a radiation worker at TMI-2 is set at 1 rem per quarter. This is done to aid in keeping individual exposures as low as reasonably achievable.

Radioactivity is also a quantity for which a system of measurement has been devised. The basic unit of radioactivity is the "curie" which corresponds to $3.7 \times 10^{10}$ (37 thousand million) radioactive decay events per second. Since this is such a large unit it is often prefixed by milli (mCi) or micro (uCi), corresponding to $10^{-3}$ (1/1000) and $10^{-6}$ (1/1,000,000) respectively.

Each type of radioactive source has a radiation hazard associated with the amount of radioactivity present. The relationship between curies and rems is dependent upon the radioisotope, the geometry of the radioactive source, and the distance from the point of measurement to the source. A pertinent example is Cs-137. The gamma radiation level at one meter from a one uCi source of Cs-137 occupying a small volume would be about 0.3 millirem per hour. The radiation level is directly proportional to the amount of radioactivity and is proportional to the inverse of the square of the distance from the source.

It is also important to distinguish between radiation and contamination. Radioactive contamination consists of small particles of radioactive material that are found on a dry surface or suspended in air or in water. The contamination, therefore, is a source of radiation.

In the working environment, measurements are taken to determine the radiation and contamination levels so that protective requirements may be established for the personnel working in that area. These requirements may include shielding, clothing, respiratory equipment, dose measuring devices, and the time that is spent in the area.